T0219137

# THE DEVELOPMENT OF MATHEMATICS THROUGHOUT THE CENTURIES

# THE DEVELOPMENT OF MATHEMATICS THROUGHOUT THE CENTURIES

## A Brief History in a Cultural Context

**Brian R. Evans**

Pace University
New York, NY

Published by John Wiley & Sons, Inc., Hoboken, New Jersey
Published simultaneously in Canada

*Library of Congress Cataloging-in-Publication Data:*

Evans, Brian R.
  The development of mathematics throughout the centuries : a brief history in a cultural context / Brian R. Evans, School of Education, Pace University, New York, NY.
       pages cm
  Includes bibliographical references and index.
  ISBN 978-1-118-85384-9 (cloth)
  1. Mathematics–History.   I. Title.
  QA21.E78 2014
  510.9–dc23
                              2013041997

10   9   8   7   6   5   4   3   2   1

# CONTENTS

# INTRODUCTION

There have been numerous books written about the history of mathematics, ranging from very brief introductions to extensive in-depth textbooks. This book is intended to provide a brief overview of the history of mathematics in a straightforward and understandable manner. By no means will this book cover the history of mathematics exhaustively or every mathematician whose work contributed to the evolution of mathematics. However, it will address major findings that influenced the development of mathematics as a coherent discipline.

Many books focus heavily on complicated mathematics, which is not easy to understand by many readers outside the discipline. The intention of this book is to present considerable mathematical content in an easy-to-digest manner. Moreover, this book will emphasize the historical foundations and background of the history of mathematics in order to provide context. There will be more mathematics content presented regarding the earlier historical periods because the mathematics discussed will be easier to understand for most readers, as opposed to its later developments. The mathematics presented in recent history would require advanced knowledge in the subject. Occasionally, however, this book will address mathematics that may assume prerequisite knowledge that may not be possessed by all readers. It is the author's belief that such an instance will rarely be encountered by most readers who have a high school background in mathematics and perhaps some college mathematics background as well. Readers who have been exposed to some calculus will have little to no trouble understanding the mathematics in this book. Hence, this book is written for a wide audience of readers. Readers with a range of basic high school or college mathematics backgrounds as well as students majoring in mathematics and high school mathematics teachers will find most of this book interesting and enjoyable. This book can be used as a textbook in a history of mathematics college class or even for pleasure reading and intellectual curiosity. Even a doctorate in the field may want to keep this book on hand as a quick reference guide. Middle school, high school, and college professors, and even elementary school teachers, may find this book useful in motivating their students to learn mathematics content through the use of mathematics history in their teaching. In order to be an effective teacher, it is important to engage students in the material. Introducing the history of mathematics can make the study of mathematics more enjoyable and help students see how mathematics developed throughout the centuries because it gives mathematics a human face. This book integrates formal history with anecdotes and legends, where appropriate, in order to make the reading more interesting. Even for the mathematical legends unlikely to be true, they have been constructed around significant historical figures and will be of interest to the reader. The reader will be notified when an anecdote or legend is merely speculative.

Another purpose of this book is to highlight the contributions made by various world cultures. The most recent books about the history of mathematics certainly have gone in this direction, but this book is intended to help the reader understand developments of mathematics around the world. Specifically, there are chapters presented on African, Egyptian, Babylonian, Chinese, Indian, Islamic, and Pre-Columbian American mathematics. It is important that students understand that mathematics has developed throughout the world and that no single culture has a monopoly on the subject. While there were remarkable mathematical developments in ancient Greece and throughout Europe in the Modern Era, these accomplishments would not have been possible without the contributions from various cultures such as Babylonian, Indian, and Islamic cultures. At the time Europe was in the "Dark Ages," particularly in mathematics and science, great accomplishments were being made in China, India, and later the Middle East. It is commonly believed that much of modern mathematics was exclusively developed in Europe, but that is not the case. As addressed in this book, many of the concepts discovered by Europeans in the 17th to 19th centuries were discovered in some form, although often but not always more limited, much earlier in China, India, and the Middle East. Rich mathematical knowledge was also produced in parts of the Pre-Columbian Americas. This book will demonstrate that Indian and Islamic mathematics were highly influential on European mathematics, which began to more fully develop in the European Renaissance and would climb to remarkable heights throughout the 17th to 19th centuries.

Students from diverse cultures and backgrounds who believe that mathematics is an exclusively white male endeavor are less likely to be engaged in mathematics class. In fact, in a study in which children were asked to draw a mathematician, many children drew the stereotypical mathematician as an older white male. Nevertheless, after bringing diverse mathematicians into the classroom to speak with the children, many children drew diverse mathematicians of different backgrounds when asked to complete the task again.

While it is important for students to perform well in mathematics for its own sake, it is essential that students perform well because it serves as the "gatekeeper" for entrance to quality colleges and high paying careers. Success in mathematics is one of the strongest variables in predicting success in life. High paying careers such as those of mathematician, actuary, statistician, and engineer all require high levels of mathematics knowledge. Strong mathematics knowledge is also necessary for personal success and in making informed life decisions, including important economic, social, and political choices. Mathematics knowledge is a necessary component of financial literacy and economic well-being, and quantitative reasoning is needed for citizens to make informed decisions in a healthy democracy. Engagement in mathematics will lead to greater learning in mathematics, and teaching students by using mathematics history as a framework can also help facilitate the necessary student motivation needed for academic success.

The history of mathematics is intriguing because it gives us a perspective of the long development of mathematical ideas throughout the ages. It allows us to glimpse into the ancient cultures and place mathematics in the context of human history and development. It helps us realize that mathematics was discovered by real people, but simultaneously mathematics reveals universal truths. Unlike other subjects that depend on subjective information, mathematics objectively tells us about the world and the universe. Once a mathematical theorem is proved correctly, it is proved for the ages. People who enjoy history can gain a positive disposition toward mathematics if the subject is presented in its historical context. Mathematics is developed and interpreted by humans throughout the ages, which makes the study of its development a worthy goal.

Mathematics can be defined as the study of patterns in the context of quantity, structure, and dimension. Throughout this book, we shall take a journey throughout time and observe how people around the world have understood these patterns of quantity, structure, and dimension around them. We shall explore mathematics in the wider historical context. This book is set up in chronological order, for the most part, beginning with the earliest known records of human mathematics in Africa. It continues with ancient Egypt and Babylon, and then covers ancient Greece to the fall of the Roman Empire. We then turn to historical developments of mathematics in ancient to medieval China and India, and on to the Islamic world. Next, the mathematics of the Pre-Columbian Americas is explored. We then go into the dark period of mathematics in Europe after the fall of the Roman Empire until the influence of Islamic mathematics on European thinking. We next visit the European Renaissance and work our way through the 17th to 19th centuries in Europe. The 20th century takes us through Europe and to North America, which took the lead in mathematics after World War II. We shall then work our way to the early 21st century. Finally, we shall explore the history of mathematics education and its development in the United States during the 20th and 21st centuries.

A final note should be made regarding the source of the information in this book. This book was written from the class notes used in a mathematics history class taught by the author. This work is considered to be a textbook of information and not an original work of historical or mathematical research. The facts have been gathered from many other sources, and none of the information presented will be an original discovery by the author, besides the occasional author speculation. Facts found in one source were checked against the same facts in other sources, when available. As a textbook written for informative purposes, references are not cited throughout the book. The Resources and Recommended Readings section at the end of this book lists all of the sources in which information was gathered. In particular, a special note should be made about the University of St. Andrew's MacTutor History of Mathematics website. This website has proven to be an invaluable source of information on the backgrounds of the numerous mathematicians found in this book. For more details than this book can provide, it is recommended that the reader explore the MacTutor website for more information. As noted throughout the text, the Internet will continue to change how we live, work, and study. Now, more than any time in the past, the reader has access to an unlimited amount of information through simply having a computer and access to the Internet. This brief book should serve as a guide through mathematics history. It places the history in context and allows the reader to have a starting point for further exploration.

# PART 1

# MATHEMATICS IN AFRICA AND THE MEDITERRANEAN REGION

# CHAPTER 1

# ANCIENT CIVILIZATION MATHEMATICS: AFRICA AS OUR BIRTHPLACE

Humanity likely originated in Africa several million years ago as it separated from the common ancestor that we share with chimpanzees. *Homo erectus* migrated to Asia, while *Homo neanderthalensis*, or the Neanderthals, migrated to Europe. However, those whom are considered to be humans today, *Homo sapiens*, actually originated in Africa 200,000 years ago and eventually replaced *H. erectus* and *H. neanderthalensis* in Asia and Europe. Our ideas of what we consider to be typically recognizable human characteristics, such as reliance on speech and abstract thought, places humans around 50,000 years ago.

It is commonly believed that humanity first evolved in the eastern regions of Africa in today's nations of Ethiopia and Kenya. Researchers have followed humanity's migration and settlements into Asia and Europe, across the Indonesian archipelago into Australia and over the Bering Straight into the Americas. Not surprisingly, mathematics developed in a similar manner. The earliest known mathematical object is the Lebombo bone found in the 1970s, which is estimated to have originated around 37,000 years ago in Swaziland in southern Africa and is named after the Lebombo Mountains. Found in the Border Cave, the Lebombo bone is a tally stick with 29 notches that

*The Development of Mathematics throughout the Centuries: A Brief History in a Cultural Context*, First Edition. Brian R. Evans.
© 2014 John Wiley & Sons, Inc. Published 2014 by John Wiley & Sons, Inc.

may have marked the lunar calendar. The Ishango bone was found in the Ishango area of the Congo in 1960 and indicates prime numbers, positive integers that have only two unique factors, between 10 and 20, and a lunar calendar like the Lebombo bone. There has been speculation that the first tally sticks indicate a lunar calendar for the purpose of predicting female menstruation cycles, which means African females may have been the first to create mathematical instruments and history recognizes them as the earliest mathematicians. The Ishango bone was found by Belgian geologist Jean de Heinzelin de Braucourt and may be about 20,000 years old. Both the Lebombo and Ishango Bones are made from the fibula, or calf bone, of a baboon. Today, the Ishango bone is on display at the Royal Belgian Institute of Natural Sciences in Brussels, Belgium.

There is also evidence of the early use of tally sticks in Europe. In 1937, the Wolf bone was found in Czechoslovakia by Karl Absolon, which may be up to 30,000 years old. It had 55 notches set in groups of five. The first 25 notches are separated from the other 30 notches by having one notch double in size from the others. Also found in Europe, the famous Lascaux Cave paintings are a collection of cave images found in southwestern France in 1940. The paintings mainly depict humans and animals, but may also have indicated a lunar calendar with 29 markings. In 1950, Willard Libby, who discovered carbon-14 dating and received the Nobel Prize for it in 1960, famously dated the Lascaux Cave paintings. Libby found that there was only 15% of carbon-14 present, as would be found in plant-based paint in 1950. The formula for exponential decay is $A = Pe^{kt}$, where $A$ is the new amount, in this case the amount of carbon-14, $P$ is the original amount, $k$ is the decay constant, $t$ is the time elapsed, and $e$ is Euler's number, which is approximately 2.7182818285. By knowing the half-life of carbon-14 to be roughly 5700 years, one could reason that since in 5700 years, we would have half of the carbon-14 present left from when an organic life-form died, in this case the planets that made up the plant-based paint, we can find the decay constant because $0.5P = Pe^{k(5700)}$. Solving for $k$ gives us $0.5 = e^{5700k}$. Taking the natural logarithm of both sides of the equation gives us $\ln(0.5) = 5700k$, and

$k$ is approximately equal to –0.0001216. Since Libby had found that only 15% of the original carbon-14 was present, we have $0.15P = Pe^{-0.0001216t}$. Solving for $t$ gives us $0.15 = e^{-0.0001216t}$. Taking the natural logarithm of both sides of the equations gives us $\ln(0.15) = -0.0001216t$, and $t$ is approximately equal to 15,600. This means the paint originated 15,600 years before 1950, which was 13,650 BCE.

The people who used the tally sticks were likely nomadic hunter–gatherers because they did not transition into an agricultural lifestyle, domesticating animals and plants, until about 10,000 years ago. It is argued that the shift from nomadic hunting–gathering to an agriculturally based system allowed for great social change for early humans. Since people became more sedentary, they gained the opportunity to begin the development of permanent settlements and stable civilizations. In fact, it is the environmental factors of good agricultural land, along with animals that are easy to domesticate, that has led noted physiologist and geographer Jared Diamond to conclude that history's most successful societies can be explained by access to these benefits.

As we shall see in later chapters, not all societies utilized the base 10 system, a system in which there are nine digits representing 1–9, a symbol for zero, and the number 10 being represented through the positioning of 1 and zero. Additionally, as we shall see, the zero took a very long time in development in human history. However, it should not be surprising that many societies, including our own modern societies throughout most of the world today, use base 10. Early humans had noticed that there was great convenience in using their fingers for counting and basic arithmetic, as many children and adults do today. Therefore, it made sense to consider the first 10 numbers as the foundation for a number system.

When numbers exceeded the number of fingers on human hands, people could turn to counting on their toes, which certainly becomes more cumbersome, but has been done in some civilizations. Another level of sophistication would be the use of the tally sticks, such as the use of the Lebombo, Ishango, and Wolf tally sticks.

For the ancient people, the next important use of mathematics was in their astronomical observations for religious purposes and tracking calendars, which would have often been connected in the celebration of holy days. Early humans, like humans of today, were fascinated by the night sky. There is mathematical regularity in the patterns found in observing the heavens. Calendar records have naturally arisen indicating each day that is designated by the rotation of the Earth, the monthly cycles of the Moon, and the yearly revolution around the Sun. We see evidence in the way the night sky influenced early mathematics in ancient Egypt, Babylon, China, India, and Greece. As humans developed stable agricultural societies, ruling and priestly classes began to develop due to societal surplus of goods. These classes would find the time to develop the mathematics of calendars for religious observances. Mathematics evolved for practical purposes such as civil affairs, including taxation, and better agricultural techniques. However, the night sky and its regularity could be considered an early birthplace of mathematics.

There is a connection between the night sky and religion. Early humans looked to the stars to find the divine. Today, this is manifested in the childhood belief in and our poetic fascination with Heaven residing in the clouds. For many cultures, the gods were found in constellations and planets. In school today, many students study the Greek names of constellations, and, even today, the planets in our solar system are named after the Greek and Roman gods. Jupiter is a Roman god, who is the greatest of gods, and equivalent to the Greek's Zeus. Saturn is the Roman equivalent to the Greek's Kronos, Neptune is the Greek's Poseidon, and Uranus is the name of a Greek god who is the father of Kronos and grandfather of Zeus. Pluto, no longer considered a planet, is the Roman name for the Greek's Hades. For the inner planets, Mercury is the Greek's Hermes, Venus is the Greek's Aphrodite, and Mars is the Greek's Ares.

The author of this book has always been fascinated by the constellations and night sky, and finds it appropriate that we can find such a connection between mathematics and the early civilizations. In a world that is ever changing as humans shape the landscape and environment around them, we can take comfort in the knowledge that when we gaze on the night sky, we are essentially viewing what ancient civilizations before us had viewed.

# CHAPTER 2

# ANCIENT EGYPTIAN MATHEMATICS: THE TIME OF THE PHARAOHS

Ancient Egyptian society was centered on the Nile River and started when the people of the region began a sedentary agricultural lifestyle using the Nile to cultivate their farmlands, which was approximately 10,000 years ago. Prior to this time, the people of the region were nomadic and lived in what is now the Sahara Desert. Desertification began around 10,000 years ago, culminating in the formation of the Sahara Desert around 4,500 years ago. Egypt became a unified state around 3150 BCE when the Upper and Lower Nile valleys joined together under the first pharaoh, Menes, which began the Early Dynastic Period. The ancient Egyptian period ends with the Roman conquest in 30 BCE, and pharaoh rule ceased with Cleopatra as the last pharaoh. The Roman Republic became the Roman Empire 3 years later when Caesar Augustus became Emperor of Rome.

The ancient Egyptians are recognized for their great accomplishments in architecture, irrigation, agriculture, art, literature, and mathematics. When considering their circumstances, it is reasonable that the Egyptians would develop their culture in the chronology that they did. When people began living a sedentary agricultural lifestyle,

*The Development of Mathematics throughout the Centuries: A Brief History in a Cultural Context*, First Edition. Brian R. Evans.
© 2014 John Wiley & Sons, Inc. Published 2014 by John Wiley & Sons, Inc.

they began to create stable villages and towns in which it was possible to build on their accomplishments and transmit knowledge to the next generation. Additionally, the wealth obtained from agriculture created need for various occupations that would require mathematical calculations for wages, architecture, and other measurements.

Egyptians needed mathematics for the creation of their calendars, which had 12 months with 30 days in each month; within each month, there were 3 weeks consisting of 10 days. The remaining 5 days were feast days at the end of the year. The Egyptians had 24 hours in a day with 12 hours for day and 12 hours for night. Twelve was a significant number for the Egyptians because they counted using the knuckles on their four fingers on one hand and each of the four fingers has three knuckles. Moreover, 12 is a convenient number given the number of factors it has.

The calendar proved useful for cultivating agriculture because the Egyptians were able to predict the best time for planting their crops and the cycles based on the flooding of the Nile River. Astronomy is directly related to calendar calculations, and as we shall see later with the Greeks, astronomy played a major role in early developments of mathematics.

A lasting legacy of Egyptian accomplishments is the pyramids, specifically the Pyramids of Giza. The Early Dynastic Period ended around 2800 BCE when the Old Kingdom Period began. It was during this time that the king became regarded as a living god and the pyramids were built. The Great Pyramid of Giza, also called the Pyramid of Khufu, was built in 2550 BCE and remains the only surviving ancient wonder of the world. It remained the tallest human-made structure until 1311 CE when the Lincoln Cathedral was built in England. The pyramids are a remaining legacy of Egyptian innovation in architecture and mathematics, and it was during this period that Egypt experienced great accomplishments in culture, art, and architecture. The Old Kingdom Period ended around 2300 BCE.

The Egyptians used the base 10 system. A number such as 432 would be expressed as four 100s, three 10s, and two 1s. However, unlike our current base 10 system, the Hindu–Arabic numeral system, the Egyptian system was neither multiplicative nor positional, but it was merely additive. That means there was no way of easily representing 800 without writing 100 eight times. In our modern system, we can represent 800 as 8 × 100, so we can see why our system is considered multiplicative. Our system is positional because we do not have to write 823 as 8 × 100 + 2 × 10 + 3 × 1, but rather we know the position of the individual digits represents hundreds, tens, and ones. For example, 823 is quite different from 382. In the first number, we know we have eight 100s, two 10s, and three 1s. In the latter number, we have three 100s, eight 10s, and two 1s. For us, it is the position of the digits that makes the difference. Using the Egyptian system, we would have to represent 823 by writing 100 eight times, 10 two times, and 1 three times. Using Egyptian symbols, we would have the following (see Table 2.1):

**TABLE 2.1   Egyptian Numerals**

| Number | Egyptian symbol | Description |
|---|---|---|
| 1 | \| | Staff |
| 10 | ⌒ | Heel bone |
| 100 | ↺ | Coil of rope |
| 1000 | ⚘ | Water lily (or lotus flower) |
| 10,000 | ↗ | Finger |
| 100,000 | 🐟 | Tadpole |
| 1,000,000 | 🧍 | Person with arms raised |

It is quite straightforward to add or subtract using the Egyptian system. For addition, let us say we have 15 + 27. We would realize that we now have three 10s and twelve 1s. In other words, we have 3 heel bones and 12 staffs. However, if we replace 10 of those staffs with a heel bone, we now have 4 heel bones and 2 staffs.

If we were to subtract two numbers, such as 14 from 31, we would immediately realize we cannot subtract the four staffs from the one staff. We would replace one of the three heel bones in 31 with 10 staffs to yield 11 − 4 = 7. Now, we only have two heel bones left from the 31 so we would have two heel bones minus one heel bone from the 14. Hence, our answer would be 17. See the illustration.

This is equivalent to the following.

The Egyptians had a very interesting method for multiplication. In order to understand their method, we must first understand a remarkable discovery made by this ancient civilization. The Egyptians knew that any integer could be written as the sum of powers of 2 without repeating any of the numbers. For example, we know that 5 = 1 + 4 and 13 = 1 + 4 + 8. The aspect of doubling may have originated in parts of Africa, south of Egypt. We shall use this rule to multiply later.

In one column, we list the powers of 2: 1, 2, 4, 8, 16, 32, . . . until we get as close to the smaller number as possible. In the right column, we start with the larger number and keep doubling. We want to find the sum of numbers in the left column that sum to the smaller number. We match those numbers with the right column and add the right column numbers together. This will be our answer. For example, $18 \times 23$ could be done as follows:

| | |
|---|---|
| 1 | 23 |
| **2** | **46** |
| 4 | 92 |
| 8 | 184 |
| **16** | **368** |

It is important to notice that $2 + 16 = 18$. Therefore, we simply need to add 46 and 368 to yield the answer for $18 \times 23$, and we get 414. This works using the distributive property because we essentially have $23 \times (2 + 16)$ to get $23 \times 2 + 23 \times 16$. Recall that the distributive property indicates that $a(b + c) = ab + ac$.

The Egyptian method for multiplication is still taught in schools today as an alternative to the traditional algorithm of multiplication that many people in the United States learned in school. *Everyday Mathematics*, a book from the University of Chicago School Mathematics Project, teaches students the Egyptian method.

The Egyptians performed division problems in the same way, for the most part. The left column would be the same as multiplication, but the right column would have consecutive doublings of the smaller number until we reach the larger number. For example, $285 \div 15$ would be as follows:

| | |
|---|---|
| **1** | **15** |
| **2** | **30** |
| 4 | 60 |
| 8 | 120 |
| **16** | **240** |

It is important to notice in the right column that $15 + 30 + 240 = 285$, the larger number. Therefore, we simply need to add 1, 2, and 16 to yield the answer for $285 \div 15$, and we get 19.

Ancient Egyptians used only unit fractions with the exception of 2/3 and 3/4, which had special symbols as well as a special symbol for 1/2. A unit fraction is a fraction with 1 in the numerator in the form $1/x$, in which $x$ is a nonzero integer. This means that 2/3 and

3/4 were the only nonunit fractions considered by the Egyptians. We could represent fractions such as 7/12 as the sum of other unit fractions. In this case, we would have 1/3 + 1/4. The procedure we can use to find the sums of unit fractions is to find all of the factors of the denominator. In our example, we could list the factors of 12: 1, 2, 3, 4, 6, and 12. We need to find the factors that sum to 7, which would be 3 and 4. Next we write 4/12 + 3/12 to yield 1/3 + 1/4. If this procedure does not work, such as in the case of 3/5, we can find an equivalent fraction such as 6/10, and then use the factors of 10 that sum to 6, which would be 5 and 1. Next we write 5/10 + 1/10 to yield 1/2 + 1/10. Finally, if all else fails, we could simply multiply the denominator by 1/2, 1/3, 1/4, 1/5, ... until we get a sum of numbers that equals the numerator. In our first example, we would have 1/3 × 12 = 4 and 1/4 × 12 = 3, and since 4 + 3 is 7, we know that our unit fractions are 1/3 and 1/4. It is this last method that was used by the Egyptians to create fraction tables.

The fraction symbol was an eye, which meant that fractions could be written with the eye on top and the denominator written under the eye. For example, the following would be the way to represent 1/3:

The following are the special symbols for 1/2, 2/3, and 3/4, respectively:

While we are on the subject of eyes, the Egyptians had a symbol called the Eye of Horus, which was a symbol of protection and was represented by the goddess Wadjet, or "Whole One." It was later associated with Horus, a god associated with the sun god Ra, who was depicted as a falcon. Each part of the eye was a unit fraction, and when taken together, they summed to 1. This is quite remarkable because even though the six unit fractions found in the eye, 1/2, 1/4, 1/8, 1/16, 1/32, and 1/64, did not really sum to 1, they came pretty close. In fact, if this series kept going it would truly be equal to 1. In other words, we would need 1/2 + 1/4 + 1/8 + 1/16 + 1/32 + 1 /64 + 1/128 + 1/256 + · · ·. This series is called a geometric series, which means that the ratio of any two consecutive terms gives us a constant ratio. In this case, that constant ratio is 1/2. For example, 1/4 divided by 1/2 is 1/2, 1/8 divided by 1/4 is 1/2, and so on. Furthermore, this sum of an infinite geometric series is $a/(1 - r)$, where $a$ is the first term and $r$ is the common ratio. In our example, we would have 1/2 divided by the quantity 1 minus 1/2, which yields 1/2 divided by 1/2 to give us 1. In Figure 2.1, we can find the first six terms in the eye. The right side of the sclera, or white of the eye, represents 1/2, the iris, or the color of the eye, and the pupil represent 1/4, the part of the eyebrow above the straight line coming out of the eye represents 1/8, the left side of the sclera represents 1/16, the curly part coming below the eye represents 1/32, and the thick straight part coming down from the eye represents 1/64.

**FIGURE 2.1**   Eye of Horus

The Egyptians also worked in geometry and were early in discovering the concept of area. They found the areas of typical mathematical shapes such as triangles and rectangles. The Egyptians had an interesting formula for finding the area of a circle, which they found by taking 8/9 of the diameter of the circle and then squaring this result. This closely compares with our own formula for the area of a circle: $A = \pi r^2$, where $A$ is the area and $r$ is the radius. $\pi$ is the ratio of the circumference to the diameter of a circle. In other words, if we measure a circle to be one meter across (the diameter), then if we wrap a measuring tape around the circle, we would have approximately 3.14159 meters. If we consider the radius to be half the diameter, then we can turn the Egyptian formula into 16/9 of the radius, and then square that result. We use 16/9 instead of 8/9 since $(8/9)d$ equals $8/9(2r)$, which equals 16/9 of the radius. Since 16/9 squared is 256/81, we can write the area as 256/81 multiplied by the radius squared. This means that the Egyptian value for $\pi$ was 256/81, which is approximately 3.160493827. This calculation is very accurate for its time and close to the actual estimate: 3.1415926536.

The ancient Egyptians developed the world's first known unit of measurement, called the cubit. The cubit is a measurement of length and is a little longer than half a meter. It was based on the length of a person's forearm to the extended fingers. It consisted of seven palms, about the height of an extended fist, which also consisted of four digits or the four fingers in a fist. This means that there were 28 digits in a cubit. A measure of volume was the hekat, which was about 1.26 gallons, or 4.78 liters. The hekat was referenced in the two famous works that will be addressed next.

The ancient Egyptians recorded their writings on papyrus, a paper-like material derived from the papyrus plant. Two of the most famous papyri come from Egypt's Middle Kingdom Period, which spanned the time from around 2200 to 1800 BCE. The two famous papyri are the *Rhind* and *Moscow Papyri*. The *Rhind Mathematical Papyrus*, also known as the *Ahmes Papyrus*, is a copy of an earlier papyrus that dates to around 1600 BCE with the original at about 1800 BCE. It was acquired in 1864 by the British Museum from Henry

Rhind, who had bought the *Rhind Papyrus* in Luxor in 1858, and the Brooklyn Museum has small fragments of the *Rhind Papyrus* today. The papyrus describes the work of the scribe Ahmes, the earliest known name in mathematics history. It had fraction tables to aid in calculation and 87 arithmetic and geometry problems, and it helps us understand the Egyptian method for multiplication. The *Egyptian Mathematical Leather Roll*, also acquired by Rhind, contains unit fraction sums. The smaller but older *Moscow Mathematical Papyrus* originated around 1800 BCE and was purchased by Vladimir Golenichev in the late 19th century in Luxor and contains 25 exercises in geometry. Today, it is housed in the Pushkin State Museum of Fine Arts in Moscow.

Before we close this chapter on ancient Egypt, we should address one of the most famous pharaohs in history, Ramesses II. The New Kingdom Period lasted from 1570 to 1070 BCE and was the time period of greatest wealth and power in ancient Egypt. This was the time of Tutankhamun, whose famous burial mask was found by Howard Carter in 1922. Thereafter, Ramesses II, who was born around 1300 BCE, reigned from 1279 BCE until his death in 1213 BCE. He is remembered for his great military victories and is considered to be the biblical pharaoh during the time of Moses and the Jewish Exodus from Egypt. Ramesses II had many wives but the best known was Nefertari. Ramesses II is the subject of the 19th-century CE English poem *Ozymandias*, written by Percy Bysshe Shelly. The poem reveals the story of a traveler who comes upon the decayed ruins of the statue of Ozymandias. On the base of the pedestal the traveler reads, "My name is Ozymandias, king of kings: Look upon my works, ye Mighty, and despair!" It is generally accepted to have a double meaning. The first meaning refers to the original intention of Ozymandias in his message to competitors, and the second refers to his message to the rulers of our time that all of their power and wealth would indeed be fleeting.

Next, we shall move on to the mathematics of ancient Mesopotamia with its center in Babylon. We shall see that the Babylonians had a system quite different from the Egyptians, but the system was just as remarkable.

# CHAPTER 3

# BABYLONIAN MATHEMATICS: THE MESOPOTAMIAN CRADLE OF CIVILIZATION

Mesopotamia is the region between the Tigris River and Euphrates River in what today is mostly modern-day Iraq, and the word "Mesopotamia" means "land between the rivers" in Greek. Mesopotamia was part of the region today called the "Fertile Crescent" because of the crescent shape of fertile land that encompasses modern-day Iraq, Syria, Lebanon, Jordan, Israel, and the Palestinian Territories. This region gave birth to the earliest human civilizations around 5000 BCE and witnessed the development of one of the first, along with Egyptian hieroglyphics, writing systems of pictographs around 3000 BCE along with the invention of the wheel. Because of the early developments of human civilization in this region, we call it, like the region around the Nile, the "Cradle of Civilization."

Mesopotamia was dominated by four empires: Sumerian, Akkadian, Babylonian, and Assyrian. The Sumerians are credited with developing the earliest writing systems, called cuneiform, around 3000 BCE. Writing took place on rectangular clay tablets using the triangular end of a reed stalk to make impressions into the soft clay. The triangular-shaped wedges that were made into the clay tablets

*The Development of Mathematics throughout the Centuries: A Brief History in a Cultural Context*, First Edition. Brian R. Evans.
© 2014 John Wiley & Sons, Inc. Published 2014 by John Wiley & Sons, Inc.

give us the name "cuneiform" after the Latin word *cuneus*, meaning "wedge" or "wedge-shaped." The Akkadians later replaced the Sumerians around 2300 BCE, but retained their writing system. After the Akkadian Period, the Babylonians flourished in the region beginning around 1700 BCE. Babylon was destroyed in 689 BCE by the ruling Assyrians. By 612 BCE, the Babylonians repelled the Assyrians, but Babylon would eventually fall under Persian rule with relative consistency from 539 BCE until the Islamic conquest around 630 CE. Sometime before the beginning of the Common Era, Babylon had ceased to be the cultural center it was in the centuries prior.

The focus of this chapter is the accomplishments of the Babylonians, who had centered their empire around the wealthy city of Babylon, which has become synonymous with Mesopotamia in general. They contributed greatly to literature and philosophy. It was in Babylon that we see the early development of formal logic and axiomatic systems. As we shall see, the Babylonians had great influence on Western thought for centuries to come and even still continue to influence us today. The Babylonian belief that the Sun and other heavenly bodies revolved around the Earth would influence Christian thought, which would persist until the 16th and 17th centuries in Europe.

Like the Egyptians, the Babylonians were successful and accomplished in many areas including irrigation, agriculture, astronomy, architecture, and mathematics. First, let us look at the Babylonian calendar. The Babylonian year began in spring and had 12 months, like our own calendar. Each month began with the crescent Moon and consisted of three 7-day weeks with the remaining 9 or 10 days representing the final week of the month. The seventh day of the first weeks were days of rest. This system influenced the Hebrew calendar and subsequently our modern calendar.

In terms of mathematical accomplishments, no other culture compares with the great mathematics developed in ancient Mesopotamia, and the Babylonians were directly influential on Greek mathematical thought. The Babylonians were the first people to use an abacus for mathematical calculations starting around 4000 years ago. However, the abacus may have been first used in Egypt. Similar devices would later be found among the Egyptians, Greeks, Chinese, and Indians. Unlike the Egyptians, who used a base 10 system that we use today, the Babylonians used a base 60, or sexagesimal, system. It is probable that 60 was chosen as a base due to the large set of numbers that divide evenly into 60 (1, 2, 3, 4, 5, 6, 10, 12, 15, 20, 30, and 60). Even today, we can see the influence of their base 60 system in our clocks with 60 seconds in a minute and 60 minutes in an hour. Consider the ease in which one can evenly divide an hour. We can have two parts of 30 minutes, three parts of 20 minutes, four parts of 15 minutes, five parts of 12 minutes, or six parts of 10 minutes. Moreover, the Babylonians even had 24 hours in a day, as the Egyptians had. Today, we are also influenced by the Babylonians in that we have 360° in a circle. Like 60, 360 can be easily broken into many factors. The Babylonians were the first to use 360° for a circle.

Remarkably, only two symbols for the numbers were needed: a horizontal wedge (▶) for 10 and a vertical wedge (▼) for 1. The truly remarkable aspect of the Babylonian system was their first use of a positional or place-value number system. It was mentioned in the previous chapter that the Egyptians did not have a positional system. Recall that in a positional system, we know that a number such as 823 can tell us that we have eight 100s, two 10s, and three 1s based on the location where the digits are positioned.

Let us begin with a simple example for numbers less than 60 that do not necessarily rely on position. It is important to note that numbers less than 60 relied only on the number of tens and ones that we have. For example, 23 gives us two 10s and three 1s, which can be represented as follows:

Now let us turn to numbers 60 and above. In the Babylonian system, 340 would be represented by five 60s and four 10s $(5 \times 60 + 40 = 340)$. This means we could symbolize this as follows. Notice the positional nature of the symbols and that we never list more than three wedges in a row or column at a time:

As another example, let us look at how the Babylonians represented 192. We have three 60s, one 10, and two 1s, represented as follows:

As a final example of this type, consider a large number such as 123,261. We first ask ourselves how many times 3600 (60²) divides into 123,261. We see that it divided 34 times. Next we look at the remainder, which is 861. We now ask how many times 60 divides evenly into 861. We see that it divides 14 times. Again, we look at the remainder, which is 21. Thus, we have $34 \times 60^2 + 14 \times 60 + 21$. In Babylonian symbols, we would have 123,261 represented as the following:

At first, there was no concept of zero for the Babylonians, just as there was no concept of zero in all ancient mathematical systems. If we were missing a power, then there would be an intentional space left. If we look at our last example without fourteen 60s, we would have the number 122,421, which can be represented as $34 \times 60^2 + 0 \times 60 + 21$. In Babylonian symbols, we would have the following:

Later the Babylonians introduced a symbol for zero to replace the empty space. This symbol was two 1 notches both written on a diagonal slant (the slant was pointing from right to left).

Since the Babylonians used a base 60 system, it is not surprising that their fractions were represented with powers of 60 in the denominator just as our modern system uses decimal denominators, which are powers of 10. For example, in our modern system, the number 0.25 means there are two-tenths and five-hundredths or $2/10 + 5/100$. In the Babylonian system, we would have in the first decimal position

1/60, 2/60, until 59/60 just as we have 1/10, 2/10, until 9/10 in our system. In the second position, we begin with 1/3600 until 3599/3600 before beginning another position, just as we have 1/100 until 99/100 in our system. It should be noted that like the absence of zero, there was no decimal point to indicate where the fractional part of the number would begin and the nonfractional part ends. Without the use of zero and a decimal point, things could get quite confusing. Numbers had to be read in context similar to the way we sometimes speak today. If you were told that something costs 695, you would know exactly what the person meant based on the item. For example, if you were buying a lunch special at a restaurant, you would know that this meant $6.95, but if you were buying a laptop computer, you know that this meant $695.00. The decimal point is not necessary when we have context.

Much of the basic operations of addition, subtraction, multiplication, and division were done very similarly to the way we perform these operations today. This is especially true of the first three operations. For division, the Babylonians essentially multiplied by the fractional reciprocal of the divisor. For example, to find 36 divided by 4, the Babylonians would write the problem as $36 \times 1/4$ using fraction tables that made the computation easier by converting the fraction into a decimal equivalent. In modern time, it would be like having 36 divided by 4 and simply multiplying $36 \times 0.25$.

The Babylonians had great accomplishments in algebra, but were hindered by their lack of algebraic symbols and notation. Everything had to be spoken or written in language. In other words, instead of having $2x + 1 = 6$ and solving for $x$, the Babylonians would have to say that we have double an unknown quantity plus one more to yield 6. To find the solution to this problem, the Babylonians would first have to say that we remove 1 from 6, and then divide by 2 to yield 3 as the unknown quantity. Remarkably, despite these problems, the Babylonians dealt with quadratics and were even able to complete the square. Here is a modern example of completing the square. Let us say we have $x^2 + 8x + 15 = 0$. First we could subtract 15 from both sides. We could cut 8 in half and square that amount to get 16. Next we could add 16 to both sides to yield $x^2 + 8x + 16 = -15 + 16$. This gives us $x^2 + 8x + 16 = 1$. We can factor the left side to yield $(x + 4)(x + 4) = 1$. This gives us $(x + 4)^2 = 1$. We can take the square root of both sides to yield $(x + 4) = 1$ or $(x + 4) = -1$. Solving both gives us $x = -3$ or $x = -5$.

The Babylonians had interesting developments in geometry. At one point, the Babylonians approximated $\pi$ to be 3.125. However, as we shall see, there is evidence that they have also used 3 for $\pi$, the same value that the Bible used as an estimate for $\pi$ in describing the building of Solomon's Temple around 1000 BCE, which was the first temple. In the first Book of Kings, chapter 7, verse 23, the diameter and circumference of a circle in Solomon's Temple is given as 10 and 30 cubits, respectively, which yields $\pi$ to have a value of 3. We can see the Babylonians at times used 3 as the value for $\pi$. This can be demonstrated in the Babylonian formula for the area of a circle. Suppose we know the measure of the circumference of a circle. The Babylonians would simply square this value and divide by 12. We can see how they did that by considering their own formula for the area of a circle: $A = \pi r^2$. We know that the circumference of a circle is $2\pi r$, which means that if we solve for $r$ we get $r = C/(2\pi)$, where $C$ is the circumference. If we substitute $C/(2\pi)$ for the $r$ in our own area formula we yield $A = (C/(2\pi))^2 \pi$, and when simplifying this, we yield $A = C^2/(4\pi)$. If we assume the Babylonians used 3 for $\pi$, we can substitute 3 for $\pi$ and we get $A = C^2/12$, which is the Babylonian formula for area of the circle.

The most famous record of Babylonian mathematics is *Plimpton 322*, written around 1800 BCE and is concerned with solving mathematical problems. It is a clay tablet that was found in the Fertile Crescent and is now in the Plimpton Collection at Columbia University in Butler Library. In 1922, George A. Plimpton acquired the tablet and later left his collection to the Columbia University Library. The number 322 indicates that it is number 322 in the collection. *Plimpton 322* contains a listing of Pythagorean triples (e.g., 3, 4, 5 and 5, 12, 13), which are triplets that have the following relationship: $a^2 + b^2 = c^2$. This formula will be familiar as the Pythagorean theorem, which will be addressed in Greek mathematics history in the next chapter. The remarkable part of the listing of the triples is that the Babylonians had knowledge of this approximately 1300 years before Pythagoras did. However, we do not know if the Babylonians made this connection to the triangle, and unlike the Greeks, we are fairly certain that the Babylonians had no way of proving this relationship to be true.

Another important artifact of Babylonian mathematics is the *Yale Babylonian Collection (YBC) 7289* tablet that originated around the same time as *Plimpton 322*. It was acquired by Yale University in 1912 and gives the oldest known representation for an approximation for the square root of 2. It gives us the square root of 2 as 1.41421296 and compared with our own 1.41421356. In addition to Columbia University and Yale University, the University of Pennsylvania houses important artifacts of ancient Babylonian mathematics. The Babylonians found square roots by a method of averaging. For example, for the square root of 10, one might guess 2, but 2 is too low. Next, we divide 10 by 2 and get 5. Thus, the answer is between 2 and 5. We find the average is $(2 + 5)/2 = 3.5$. So let us divide 10 by 3.5 and we yield about 2.9. We take the average again with the previous estimate and yield $(3.5 + 2.9)/2 = 3.2$. We can keep repeating this process as we get closer to approximately 3.162, a good approximation for the square root of 10.

We should say something of the remarkable work done by Otto Neugebauer, who was a 20th-century Austrian mathematician who specialized in the history of Egyptian and Babylonian mathematics with emphasis on studying Babylonian clay tablets. His major contribution is the expansion of our knowledge about Babylonian mathematics. Neugebauer discovered that the Babylonians were much more advanced than historians had previously realized up until his time. Neugebauer provided us with a large number of discoveries about Babylonian mathematics.

Before we close the chapter on ancient Babylon, we should mention one of its most prominent rulers, Nebuchadnezzar II, born around 634 BCE and who ruled Babylon from around 605 to 562 BCE. The legend says that his wife missed the greenery of her home in Persia, so in an effort to please her, Nebuchadnezzar II created the Hanging Gardens of Babylon, one of the seven wonders of the ancient world. In the Bible, he is known for conquering Judah, destroying Solomon's Temple around 590 BCE, and sending the Jews into exile.

Next we shall examine a great heritage of mathematics, the mathematics of the Greeks. It is in Greek mathematics that we truly see great advancements and, for the first time, witness the development of the idea of theoretical mathematics, which was supported by proof.

# CHAPTER 4

# GREEK MATHEMATICS: THE ARCHAIC PERIOD

Ancient Greece can be divided into three periods: Archaic, Classical, and Hellenistic periods. The Archaic Period lasted from about 800 to 500 BCE and was followed by the Classical Period from about 500 to 323 BCE, culminating with the death of Alexander the Great. The Hellenistic Period lasted from 323 to 146 BCE, ending with the conquest of Greece by the Romans. However, even after the fall of Greece to the Romans, Greek culture continued to flourish and was accepted by and integrated into Roman culture. Greek philosophy and mathematics continued to develop even after the end of ancient Greece. Therefore, a fourth period will be considered that could be called the Roman/Byzantine Period, which lasted from 146 BCE until the fall of Rome in 476 CE or the closing of Plato's Academy in 529 CE by Byzantine Emperor Justinian I. However, Greek philosophical thought was challenged by Christianity in the first several centuries of the Common Era. Roman Emperor Constantine converted to Christianity in 312 CE, and in 391 CE, Christianity became the official religion of the Roman Empire, which furthered the decline of Greek thought. These next three chapters are concentrated on the development of

*The Development of Mathematics throughout the Centuries: A Brief History in a Cultural Context*, First Edition. Brian R. Evans.
© 2014 John Wiley & Sons, Inc. Published 2014 by John Wiley & Sons, Inc.

Greek mathematics divided such that this chapter focuses on the Archaic Period, the next chapter focuses on the Classical Period, and the following chapter focuses on the Hellenistic and Roman/Byzantine periods.

The Archaic Period is characterized by the art and sculptures that the Greeks produced during that period. This was the time during which epic poet Homer likely wrote (but perhaps had written earlier) the *Iliad* and the *Odyssey*, which were consecutive works about the Trojan War and Ulysses. Greece was a collection of city-states that had natural borders separated by mountains and the Aegean Sea, Ionian Sea, and the Sea of Crete in the Mediterranean. During this period, we can see the development of the Greek alphabet, the first Olympic Games, and one of the world's earliest uses of coins as currency. It was at the end of the Archaic Period when Athens developed into the world's first democracy. The Greek city-states saw remarkable growth in wealth during the Archaic Period, and many historians attribute this to their unique geographical positions. Due to the mountains and seas surrounding the Greek city-states, agriculture was difficult and the city-states had to engage in seafaring and trade. This exposed the Greeks not only to foreign ideas but also to the many goods and services that could be made available from other city-states and neighboring regions. Just as immigration brought great prosperity to nations such as the United States in the 19th and 20th centuries by attracting intelligent and innovative people from around the world, immigration to the Greek city-states brought talented people to Greece that would lead to further advancement and wealth. With this surplus wealth, Greek society could support a class of thinkers who would be free to use their time to develop philosophical and mathematical thought.

During this time, we see the development of a new kind of mathematics. The Greeks were highly influenced by Egyptian and Babylonian mathematics, but the Greeks differed in a very important way. The Greeks were the first to move beyond the use of practical and applied mathematics toward a theoretical approach to the subject. Prior to this time, mathematics was developed only for practical purposes, but the Greeks were not interested only in mathematical applications. They studied mathematics for the beauty and truth of their discoveries, and they were the first to require formal proof for their claims. They began with an axiomatic system, or statements that were obvious and assumed to be true, and would use logic to derive theorems, which are statements that have been proven to be true. The Greeks would use their axioms and theorems to build more theorems so that their mathematics would be built from a solid foundation and would continue to advance. This was the truly magnificent contribution, among many others, originating in ancient Greek thought. Given the influence of Egyptian and Babylonian mathematics on Greek mathematics, we can trace the lineage of our modern mathematics directly back to these three cultures.

The Greeks were primarily concerned with geometric developments. However, it would be good to address the Greek number system. The earliest Greek numbers used are called the Attic numerals and predate the Greek alphabet. They were simpler than the more sophisticated system called Ionian Greek numerals that were implemented by the Greeks using lowercase alphabet letters to represent numbers. The Attic numerals were as follows: $1 = |$ , $5 = \Pi$, $10 = \Delta$, $100 = H$, $1000 = X$, and $10{,}000 = M$. The Attic numerals were used from about 700 BCE until about 400 BCE when they were replaced with the Ionian Greek numerals (Table 4.1). The Greek number system was a base 10 system.

Thales is considered to be the first Greek philosopher and mathematician. He lived from approximately 620 to 550 BCE and was from

**TABLE 4.1   Ionian Greek Numerals**

| Number | Name | Symbol | Number | Name | Symbol | Number | Name | Symbol |
|---|---|---|---|---|---|---|---|---|
| 1 | Alpha | α | 10 | Iota | ι | 100 | Rho | ρ |
| 2 | Beta | β | 20 | Kappa | κ | 200 | Sigma | σ |
| 3 | Gamma | γ | 30 | Lambda | λ | 300 | Tau | τ |
| 4 | Delta | δ | 40 | Mu | μ | 400 | Upsilon | υ |
| 5 | Epsilon | ε | 50 | Nu | ν | 500 | Phi | φ |
| 6 | Digamma | σ | 60 | Xi | ξ | 600 | Chi | χ |
| 7 | Zeta | ζ | 70 | Omicron | ο | 700 | Psi | ψ |
| 8 | Eta | η | 80 | Pi | π | 800 | Omega | ω |
| 9 | Theta | θ | 90 | Qoppa | ϙ | 900 | Sampi | T |

Miletus, which is an island off the western coast of modern-day Turkey. He is considered to be the "Father of Science" and was one of the Seven Sages of Greece. Along with Pythagoras, Socrates, and several others, Thales is credited with encouraging people to "know thyself," which was inscribed on the Temple of Apollo at Delphi. Thales is considered the first person to introduce skepticism and rational thought into Greek philosophy because he sought to explain the world through rational inquiry rather than reliance on unproven mythology. He was an early proponent of naturalism, which states that reality can be explained through natural means and precludes supernatural causes. After traveling to Egypt and learning about Egyptian geometry, Thales was the first to introduce geometry into Greece, the branch of mathematics that would dominate Greek mathematics for centuries. Thales calculated the height of the pyramids by using similar triangles while he was in Egypt. He noticed that at a certain time of day his own shadow was equal to his height, which allowed him to measure the shadow of a pyramid and conclude that this shadow equaled the pyramid's height. In addition to similar triangles, it is believed that Thales also knew how to find congruent triangles using, for example, the angle–side–angle postulate.

Thales proved a theorem that shows that vertical angles, such as angles 1 and 2 in the illustration below, are congruent. He essentially showed that since a straight line has 180° and each pair of vertical angles shares a common adjacent angle (angle 3 in the illustration), we have angle 1 + angle 3 = 180° and angle 2 + angle 3 = 180°. This means that angles 1 and 2 are congruent.

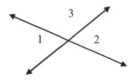

Thales proved that the base angles of an isosceles triangle are congruent. An isosceles triangle is a triangle with at least two equal sides. In the Middle Ages, this was called *pons asinorum*, which is Latin for "bridge of asses" or "bridge of fools," and was used to separate the students who would continue to study mathematics from those who would not. If we consider an isosceles triangle and drop a median line (which cuts the base of the triangle in half), we realize that based on the definition of an isosceles triangle, we have the left and right sides of the large triangle congruent. Since the base was cut in half, we have the two new bases congruent. Finally, since we dropped a median, through the reflective property, the median is equal to itself. Hence, we created two triangles that are congruent, and we can conclude that the corresponding base angles are congruent.

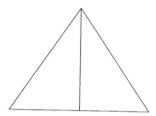

The next notable mathematician from this period is one whom many high school students of mathematics can easily recognize based on his famous theorem: $a^2 + b^2 = c^2$, where $a$ and $b$ are legs and $c$ is the hypotenuse of a right triangle. This mathematician is Pythagoras of Samos, who lived from around 570 to 475 BCE. Samos, like Miletus, is an island off the western coast of modern-day Turkey. In 535 BCE, Pythagoras, like Thales before him under whom Pythagoras probably had studied, traveled to Egypt where he learned about Egyptian mathematics. While Pythagoras was still in Egypt, the Persians invaded. Pythagoras was taken as prisoner and brought back to Babylon. After his release, he returned to Samos to start a school called the Semicircle. By this time, Samos was also under Persian control.

It is important to note that even though we attach Pythagoras' name to the famous right triangle theorem, it is likely that Pythagoras encountered Pythagorean triples in his learning of Egyptian and Babylonian mathematics. However, we name the theorem after Pythagoras for a good reason. It was merely a conjecture before the time of Pythagoras, since a theorem can be called a theorem only once it has been proven. The Pythagorean theorem has been proven in hundreds of different ways since this time, and even James Garfield, the 20th president of the United States, provided his own proof.

We shall present one proof of the Pythagorean theorem here, and another in Chapter 8. We can draw the figure (see below) so that the big square is, in fact, really a square, and create four identical triangles inside. The reader can label the small leg of the triangle $a$, the larger leg $b$, and the hypotenuse $c$. Thus, the area of one triangle is $(1/2)ab$. Since there are four triangles, we have a total area of $2ab$. The sides of the large square are $(a + b)$ each, so the area of the large square is $(a + b)(a + b) = a^2 + 2ab + b^2$. The area of the small interior square is $c^2$. The area of the big square equals the area of the small square plus the four triangles, which means $a^2 + 2ab + b^2 = c^2 + 2ab$ or $a^2 + b^2 = c^2$. The only aspect of this proof that is questionable is our assuming that the small interior square is really a square. We can consider that the sum of the angles of a triangle is 180°. This means that the sum of the non-right angles in a right triangle is 90°. If we focus on any corner of the small interior square, we see that there is a straight line, or 180°, that contains two angles that should sum to 90°. This means that the small interior square really is a square since its angle must be 90°.

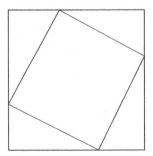

In 530 BCE, Pythagoras moved to Croton in southern Italy to start his religious sect that was known as the Pythagoreans. The Pythagoreans practiced an ascetic lifestyle; the members of the inner circle were vegetarians and eschewed all possessions. All property was considered to be community property. The term "vegetarian" was adopted in the English language in 1847 by Joseph Brotherton in the formation of the Vegetarian Society of the United Kingdom. However, until 1847, vegetarians were called Pythagoreans (and hence the

author of this book can be considered a "Pythagorean"). The Pythagoreans emphasized ethics and found virtue in honesty, friendship, and charity. They functioned as a secret society and used the pentagram as their symbol, a symbol in which much mathematics can be found. It is important to note that women were not excluded from this group and were considered equal to men. It has been recorded that nearly 30 women may have studied mathematics as Pythagoreans. One student was Theano, who may have been the wife of Pythagoras. Theano continued the school after the death of Pythagoras. It is possible that she may have conducted work with the golden ratio, an idea discussed later in this chapter. It is likely that the daughter of Pythagoras and Theano was Damo, who may have published her father's work after his death.

Aristotle had considered the Pythagoreans to be the first group to have taken up the study of mathematics. The Pythagoreans generally could be split into two groups based on their roles and contributions: those who contributed to the development of mathematics and those who focused on religion. The Pythagoreans believed that everything could be represented as numbers and they said that all things are numbers. They were famous for attributing many things in the world to a prescribed natural or counting number. Even today, we agree that all natural things can be analyzed quantitatively. The number 10 had special significance to the Pythagoreans because 10 can be expressed as the sum of 1, 2, 3, and 4. The number 1 represented no dimensions such as a point, the number 2 represented one dimension of two points to make a line, the number 3 represented two dimensions such a triangle in a plane, and the number 4 represented three dimensions such as the tetrahedron.

The Pythagoreans were interested in even and odd numbers, and considered even numbers to be female and odd numbers to be male. It is interesting that we can make certain claims about operations with even and odd integers. If we represent an even number as $2m$ and an odd number as $2n + 1$, we could conclude that an even number times an odd number is an even number because $(2m) \times (2n + 1) = 4mn + 2m = 2(mn + m)$. We notice that $2(mn + m)$ is another even number.

Note that we represented the odd number as $2n + 1$ because we did not want to imply that the odd number was the number directly after our even number. Other such operations could be represented in a similar manner.

Pythagoras observed that if we look at the proper factors of a number, the sum of the factors either are less than the number, equal to the number, or greater than the number. Proper factors are all of the factors except for the number itself. For example, the proper factors of 8 are 1, 2, and 4. Since $1 + 2 + 4 = 7$ is less than 8, then we could call 8 deficient. The proper factors of 12 are 1, 2, 3, 4, and 6. Since those numbers sum to 16, we could call 12 abundant. The proper factors of 6 are 1, 2, and 3, which happen to sum to 6. We could call 6 perfect. We also know that 28 is a perfect number because $1 + 2 + 4 + 7 + 14 = 28$. The next perfect number is 496. Perfect numbers become more difficult to find after this and will be much larger. Amicable numbers are pairs of numbers in which the sum of the proper factors of each equals the other number. The first pair of amicable numbers is 220 and 284 because the factors of 220 sum to 284: 1, 2, 4, 5, 10, 11, 20, 22, 44, 55, and 110. The factors of 284 sum to 220: 1, 2, 4, 71, and 142.

To be more precise regarding the statement of belief that all things are number, the Pythagoreans believed everything could be represented by rational numbers. If we consider a right triangle with legs both equal to 1, then we can use the Pythagorean theorem to find the hypotenuse, which would give us $\sqrt{2}$, an irrational number. An irrational number is a number that cannot be expressed as the ratio of two integers. As a decimal expansion, irrational numbers go on infinitely with no repeated pattern. At first, this was outside the world view for the Pythagoreans. A legend says that when one Pythagorean, Hippasus of Metapontum, discovered irrational numbers, he was drowned at sea by his fellow Pythagoreans to keep this idea quiet. Incidentally, recall that the Babylonians already had a concept for $\sqrt{2}$, even if they did not fully understand it at the level in which we do today. Perhaps the Pythagoreans were not aware of this Babylonian development.

Euclid, who is discussed in Chapter 7, provided a nice proof to show that $\sqrt{2}$ is irrational, although it is generally agreed that he was not the originator of the proof. Aristotle is known to have hinted at this proof by stating that if $\sqrt{2}$ were rational, all odd numbers would really be even. Euclid presented a proof by contradiction, or *reductio ad absurdum*, translated to "reduction to the absurd," which meant that we were to assume the opposite of what we were trying to prove and show that this leads to a contradiction. Hence, our original assertion must have been true. Euclid assumed that $\sqrt{2}$ was rational, which would mean $\sqrt{2} = a/b$, where $a/b$ is a fraction in simplest form. Next, we square both sides to yield $2 = a^2/b^2$. This gives us $2b^2 = a^2$. This means that since $a^2$ is even, then $a$ must be even. Since $a$ is even, we can write $a$ as $2m$. Substituting this gives us $2b^2 = (2m)^2$, which means $2b^2 = 4m^2$. This simplifies to $b^2 = 2m$, which means that $b^2$ is even and thus $b$ is even. Since both $a$ and $b$ are even, we have a contradiction because this would mean the original fraction was not in simplest form as was claimed. Our claim is now proved and we can end the proof with the letters Q.E.D., which is an abbreviation for the Latin *quad erat demonstratum*, meaning it has been shown or demonstrated.

One final point to the proof should be noted. We assumed that if $a^2$ was even, then $a$ was also even. Since $a^2$ is $a \times a$, and we know that this product is divisible by 2, then we know that one of the $a$'s must be divisible by 2 (in fact, both are since they are both the same number). This means that if $a^2$ is even, $a$ must be as well.

Many of the contributions from Pythagoras may have come from either Pythagoras or his followers. Another example that may have originated with Pythagoras or perhaps his followers was showing that there are 180° in every triangle by demonstrating that three interior angles of any triangle can be rearranged so that they fall perfectly on a straight line. However, this may have been done first by Thales. There are no remaining written documents of their achievements, only references to their achievements from later writers. Besides direct contributions to mathematics in general, it is believed that Pythagoras and his followers contributed greatly to music theory.

Using mathematics, Pythagoras was able to understand that by holding a string in the center of the string, he would raise the note one octave higher. Hence, we see that even music can be analyzed mathematically. Everything truly is numbers. In addition to all of his accomplishments mentioned earlier, Pythagoras contributed to astronomy and medicine.

It is generally believed that Pythagoras and the Pythagoreans were the first to develop the idea of a golden ratio, but it was more fully expanded by Euclid over 200 years later. The golden ratio can be found in the Pythagorean symbol of the pentagram. The golden ratio is the ratio of a large line segment ($b$) to a small line segment ($a$) such that the ratio is equal to the sum of the two line segments ($a+b$) to the large line segment ($b$).

$$a \qquad\qquad b$$

We can take the small line segment to be $a$ and the large line segment to be $b$. This means we have $b/a = (a + b)/b$ or $b/a = a/b + 1$. If we let $x = b/a$, then we have $x = 1/x + 1$. Multiply both sides by $x$ and we have $x^2 = 1 + x$, which becomes $x^2 - x - 1 = 0$. If we apply the quadratic formula, which is addressed in a later chapter, we get $x = 1/2 + \sqrt{5}/2$ or $1/2 - \sqrt{5}/2$. If we focus only on the first root, we see that this value is approximately the irrational number 1.61803. This is the golden ratio and is named by the Greek letter phi, or $\varphi$. If we look at the absolute value of the negative root, we have 0.61803.

The golden ratio is considered by many to be a very aesthetically pleasing ratio. It often takes the form of a golden rectangle, which is a rectangle of the proportion $1 : \varphi$. A rectangle that is close to golden would be a $3 \times 5$ inch index card, a television screen, or the front of a cereal box. It is said that the Greek Parthenon was built with the golden ratio in mind. Some have claimed that the most pleasing human proportions in face and body have golden ratio relationships. Leonardo da Vinci called the golden ratio the divine proportion and it can be seen in his *Mona Lisa* and *Vitruvian Man*. The golden spiral, which is based on the golden ratio, appears often in nature, such as in the pine cone and nautilus shell.

In the next chapter, we shall continue our study in the Classical Greek Period, in which we shall see great mathematical accomplishments. This period is commonly referred to as the Classical Greek Period due to the great classical contributions from this time period in many areas of study. In the following chapter, we shall also look at the development of Greek mathematics during the Hellenistic Period and the time during the Roman and Byzantine empires.

# CHAPTER 5

# GREEK MATHEMATICS: THE CLASSICAL PERIOD

The Classical Period began around 500 BCE and continued until the death of Alexander the Great in 323 BCE. This is the time of Socrates, Plato, and Aristotle. It is a time in which we see great Greek accomplishments in philosophy, art, architecture, politics, literature, and of course mathematics. This period generally began with the introduction of democracy to Athens in 510 BCE when the Spartans helped the Athenians overthrow the tyrant Hippias. It continued with the Athenian victories against the Persian invaders from 490 to 480 BCE, which led to a century of Athenian dominance in the region. During the Persian Wars, the famous Battle of Marathon took place; today it is the name for the 26.2-mile race. Pericles led Athens during its Golden Age, which lasted between the Persian Wars and the Peloponnesian Wars, and it was during the Golden Age of Athens that the Parthenon was built. The Peloponnesian Wars were between the Peloponnesian League, led by Sparta, and the Delian League, led by Athens. In 404 BCE, the Peloponnesian Wars left the Athenians defeated with temporary suspension of Athenian democracy.

*The Development of Mathematics throughout the Centuries: A Brief History in a Cultural Context*, First Edition. Brian R. Evans.
© 2014 John Wiley & Sons, Inc. Published 2014 by John Wiley & Sons, Inc.

In 371 BCE, Thebes gained victory over the other Greek city-states. This conquest was short-lived as it was soon replaced by Macedonia as the dominant force in the region around 350 BCE, during the rule of Phillip of Macedonia, father of Alexander the Great. Alexander was born in 356 BCE, and when he was only 20 years old, he inherited the throne from his father, who had been assassinated in 336 BCE. Alexander continued his father's conquest of all of Greece through military force and diplomacy. In 334 BCE, Alexander expanded his empire through the acquisition of the Persian Empire, Egypt, and the northwestern parts of India. In Egypt, Alexander was told by the oracle that he was from divine origin, and thereafter he founded the city of Alexandria. The Library of Alexandria, the most significant library and place of learning at the time, was later founded by a student of Aristotle. It is probable that Alexander died from disease in 323 BCE in the city of Babylon.

Socrates and his student Plato, as well as Plato's student Aristotle, were philosophers credited with the foundation of Western philosophy. All three were said to have been Pythagoreans, or vegetarians, but the strongest evidence is that only Plato was a vegetarian and a follower of Pythagoras; the others may not have been.

Socrates was born during the Classical Period in 469 BCE. There is no evidence of his own writings, but his philosophy was recorded by Plato, in which Socrates used his Socratic method to generate knowledge. The Socratic method poses challenging questions to a student so that the student is forced to formulate intelligent answers, often refining his or her statements through various challenges. After his criticism of Athenian leadership, Socrates was forced to commit suicide by the state in 399 BCE for allegedly corrupting the minds of the youth of Athens and not believing in the gods of the state.

Plato, a student of Socrates, was born sometime after 430 BCE in Athens. In addition to philosophy, Plato studied ethics and mathematics. Plato's best-known work is the *Republic*, in which he expresses the dialog of Socrates about politics, justice, and the nature of reality. In the *Republic*, Plato recommended that education revolve around four studies with a strong mathematics emphasis: arithmetic, geometry, astronomy, and music. In the Middle Ages, this was called the quadrivium, or four ways, and was four of the seven parts of a liberal arts education. The other three parts were logic, grammar, and rhetoric, often called the trivium. Around 387 BCE, Plato founded the Academy, in which it is believed that the sign above the door read that no one who was ignorant of geometry could enter. Some of the five Platonic solids, named after Plato because of his writings on the topic, were known to Pythagoras. A Platonic solid is a three-dimensional shape (specifically, a convex polyhedron) consisting of regular polygons. The five Platonic solids are tetrahedron (consists of four triangles), cube (consists of six squares), octahedron (consists of eight triangles), dodecahedron (consists of 12 pentagons), and icosahedron (consists of 20 triangles). Each solid corresponded to an element: fire, earth, air, ether, and water, respectively. Interestingly, Plato believed that mathematics existed separately from the real world. For example, we do not need any collection of three physical objects to discuss the idea of 3. The number 3 exists abstractly in a perfect form not dependent on physical reality. Shapes can be thought of in a similar manner. Regardless of how hard we try, we are unable to construct a perfect circle or perfect square because there will always be a flaw, even if it is minuscule. We can, however, conceive of a perfect circle or perfect square *a priori* in our minds. Plato died around 350 BCE, but his student continued to expand classical Greek philosophy.

Aristotle, born 384 BCE, came to Athens at 18 years old to study at Plato's Academy, and he would eventually teach at his own school, the Lyceum. He was a teacher of Alexander the Great, whose death ended the Classical Period. Additionally, Aristotle was a prolific writer on many subjects. Aristotle's mathematical contributions were primarily in the form of deductive logic. Aristotle felt that reason was the path to the truth and employed a scientific method in his observations. Aristotle believed that it was important to begin with axioms that were assumed to be true and then to use logic to deduce new knowledge. However, even if one were to choose false axioms, as long as one could logically deduce statements generated by these axioms, then the argument would still be valid, but not sound. A classic example of logical reasoning is the following:

> If I live in Paris, then I live in France.
> I live in Paris.
> Therefore, I live in France.

This is a valid argument because it is impossible to assert the first two premises and yet deny the conclusion. It may turn out that one of the first two premises is wrong. The argument would remain valid, but would not be sound. For example,

> If I live in Paris, then I live in England.
> I live in Paris.
> Therefore, I live in England.

While this is a valid argument, it is not sound because clearly the first premise if false. However, the logic is valid, but due to the false premise, we arrive at a false conclusion. This is an example of modus ponens, which is Latin for "mode that affirms."

Thomas Aquinas, a Catholic philosopher in the 13th century CE, referred to Aristotle as "the philosopher" and applied Aristotle's reasoning methods to Christian thought. Until that time, many Christian thinkers had rejected non-Christian philosophers. Aristotle died in 322 BCE, a year after Alexander the Great, and at the beginning of the Hellenistic Period.

An important mathematician during the Classical Period was Hippocrates of Chios, who is not to be confused with the Hippocrates after whom the Hippocratic Oath is named in medicine. Hippocrates was born in 470 BCE in Chios, which is off the coast of modern-day Turkey. He may have devised the *reductio ad absurdum* that we used in the previous chapter. Much like Euclid did over a hundred years later based on the influence of Hippocrates, Hippocrates summarized Greek mathematical accomplishments up until his time in his book,

*Stoicheia Elements*. Hippocrates worked on the ancient problem of "doubling the cube." This means we are to construct a cube that is double in volume of a cube in which the three dimensions are each 1 unit in length, using a compass and a straight-edge. This has been shown to be impossible.

Another major problem for Greek mathematicians was squaring the circle. By this we mean to construct a square that has the same area of a given circle using only a compass and a straight-edge. Hippocrates was one of the earlier mathematicians to attempt this problem. This was proven to be impossible in 1882 CE because $\pi$ is a transcendental number, which means that it cannot be the root of any polynomial with rational coefficients. Consider that the area of a circle is $\pi r^2$, which involves the irrational number $\pi$. Therefore, in order to square the circle, our circle would need to have a side that is $\sqrt{\pi}r$ units so that when we find the area of our square, we yield $\pi r^2$ units. The problem we find for ourselves is taking the square root of $\pi$. Hippocrates worked with circles in order to attempt this problem of squaring the circle. He calculated the area of part of intersecting circles called a lune, which has the appearance of a phase of the Moon, hence the name. A third famous insolvable problem during this time was to trisect an angle using only the compass and straight-edge. Hippocrates died in 410 BCE.

The philosopher Zeno of Elea was born around the same time as Hippocrates. Elea is a town located in southern Italy. Zeno is best known for his paradoxes, many of which follow the same paradoxical idea that in order to get from point A to point B, one must pass the midpoint, which we can call C. However, to get from A to C we need to pass the midpoint for A and C, which we shall call D. This reasoning continues *ad infinitum* until we realize that no motion is possible. But yet motion is possible from our experience. This is the paradox and is called a dichotomy. A similar paradox is the story of Achilles and the Tortoise. The paradox claimed that a fast runner could never overtake a slow runner if the slow runner has any head start since the fast runner will always have to reach the slow runner's spot before passing. By this point the slow runner will have gone a little farther, even if it is less distance covered than the distance covered by the fast runner in the same period of time. Zeno died around 430 BCE.

Another important thinker in the Classical Period is Democritus. He was born in Thrace around 460 BCE and died around 370 BCE. He probably knew Hippocrates, Socrates, Plato, and Aristotle. Democritus was born into a wealthy family and traveled extensively for his learning. It was said that Plato severely disliked Democritus and wished for all of his books to be burned. His biggest contribution to science is his development of atomic theory, that is, that everything is composed of atoms, which he believed to be indivisible. Hence, he is called the "Father of Modern Science." His mathematical contributions included his discovery of the relationships between a cone and cylinder, and a pyramid and prism. He found that the former had one-third the volume of the latter if they had the same base and an equal height. This would later influence the greatest mathematician of antiquity, Archimedes. Democritus is sometimes referred to as the "laughing philosopher" for his propensity to mock those he felt were not as intelligent as he was. However, he may have been known by this title because of his own happiness. He was known for his simple lifestyle with interest primarily in his studies, and he is famous for possibly being an atheist. The legend says that Democritus blinded himself so that he would not be distracted in his studies. Although this is probably not true, he did go blind later in life. Democritus' belief in atoms would be adopted by Epicurus, who lived from around 340 to 270 BCE. Epicurus was possibly a Pythagorean in the sense that he may have been a vegetarian, and he advocated simple living. His philosophy was optimistic concerning living a happy life, and he was an early thinker to propose a scientific method that suggested beliefs should be based on direct observation and logical thinking. Epicurus may have been an atheist as well, and is famous for addressing the problem of evil. He said that if God wishes to prevent evil but cannot, then he is not omnipotent. Moreover, God is not omnibenevolent if he can but does not wish to prevent evil. If God wishes and is able to prevent evil, then there is the question of why evil exists. If he neither wishes nor is able to prevent evil, then he should not be called "God."

While on the topic of religion, Strato of Lampsacus should be mentioned, given his scientific approach. Strato was born around 335 BCE in Lampsacus, Greece. He believed the idea of god was not necessary to explain natural phenomenon in the universe and that observation and research could provide all explanation. This essentially means that everything in the universe could be examined using a scientific method of analysis and that nothing requires supernatural explanation, which makes him an early naturalist. Strato was similar in some ways to Democritus in his belief of the composition of matter. Strato died around 270 BCE.

In closing this chapter, a final note can be made about education in ancient Greece. Education was highly valued by the Greeks; however, there were two major schools of thought on what education should be in ancient Greece. The Athenian system was oriented toward the quadrivium of arithmetic, geometry, astronomy, and music, as previously mentioned, which was highly influenced by Socrates, Plato, and Aristotle. Other areas of study included philosophy, architecture, art, poetry, and physical education. The more warlike Spartan system involved education in preparation for war. Physical education was a major focus along with the art of fighting.

In the next chapter, we follow the development of Greek mathematics into the Hellenistic Period, which begins at the time of the death of Alexander the Great in 323 BCE. We then continue our journey through the Roman and Byzantine empires.

# CHAPTER 6

# GREEK MATHEMATICS: THE HELLENISTIC AND ROMAN/ BYZANTINE PERIODS

The Hellenistic Period, which is differentiated from the Archaic and Classical periods, or Hellenic Greece, began with the death of Alexander the Great in 323 BCE in Babylon. Aristotle, the last of the great three Greek philosophers of the Classical Period, had died 1 year prior in 322 BCE. After Alexander's death, the Greek Empire that he created fractured into various kingdoms led by Alexander's quarrelling generals known as the Diadochi. When asked who his successor should be, Alexander's dying words were allegedly that the strongest should rule. The word for "strongest" was similar to the name of one of Alexander's generals, which means we are not certain of his intentions. In the ensuing struggle for power, the empire broke into various regional kingdoms including the Ptolemaic Kingdom in Egypt, Seleucid Kingdom in Mesopotamia, and Antigonid Kingdom in Macedonia, among others. However, during this time period, the Greek Empire was at its zenith. The Hellenistic Period continued until the Roman conquest of Greece in 146 BCE, but because the Romans respected and adopted Greek culture, it could be argued that the spirit of Hellenistic Greece continued until the rise of Christianity in Rome.

*The Development of Mathematics throughout the Centuries: A Brief History in a Cultural Context*, First Edition. Brian R. Evans.

During the Hellenistic Period, there were great accomplishments in Greek mathematics. Probably born around the time of Alexander's death is a mathematician that we know very little about, Euclid of Alexandria. Euclid may have attended Plato's Academy and it is known that he taught at the Library of Alexandria, which was considered the center of the Western academic world at that time. Euclid may not have actually contributed a great deal to mathematics in terms of making new discoveries. Rather, he is remembered for being the compiler or editor of the great works up until his time, and is often considered to be the "Father of Geometry." Some of Euclid's proofs were original and developed by him, including his own proof of the Pythagorean theorem. Similar to the Pythagorean theorem, alternative proofs had been already developed by other mathematicians. Euclid was very successful in that he wrote one of the most important books in human history, and it is very easily the most important book in the history of mathematics. His book, the *Elements*, has appeared in more editions than any other book besides the Bible. It was one of the first mathematical books printed after the invention of the printing press. Even today's modern high school geometry book is typically modeled after the *Elements*. The *Elements* formed the foundation for nearly all mathematicians to follow over the next 2000 years, and is regarded as the greatest mathematics book ever written. Because this book summarized previous work so well, most of the original copies of works done prior to the publication of the *Elements* have been lost. Written about 300 BCE, the *Elements* is actually a collection of 13 books. The first six books explore plane, or two-dimensional, geometry, and it is here that the work of Pythagoras is presented. The next three books address number theory, followed by the book that presents irrational numbers. The last three books explore three-dimensional geometry, and the final book is focused on the Platonic solids.

An important contribution of Euclid's work is the introduction of definitions, such as defining space as the set of all points. At the foundation of Euclid's work are undefined terms such as points, lines, and planes. From undefined and defined terms, we can state postulates, such as Euclid's famous five postulates. For postulate 5, we shall use the definition from the 18th-century Scottish mathematician John Playfair.

1. A straight line segment can be drawn by joining two points.
2. A straight line segment can be drawn indefinitely to make a straight line.
3. A circle can be drawn from a center point to any constant distance called the radius.
4. All right angles are congruent.
5. For a line and a point not on the line, there is only one other line that could contain the point that is parallel to the first line.

It is the last postulate that has caused great contention for 2000 years. Many mathematicians sought to prove the "fifth postulate," famously referred to as the "parallel postulate," so that it would be considered a theorem instead of a postulate. However, no attempt to prove the fifth postulate has ever worked by only assuming Euclid's other postulates. Not assuming the parallel postulate gave rise to non-Euclidean geometries in the 19th century.

Euclid used undefined terms, defined terms, and his postulates, or axiomatic system, to prove theorems. He then proceeded to use the undefined terms, defined terms, postulates, and previously proved theorems to build new theorems. Hence, we see that Euclid's biggest contribution was bringing Greek mathematical knowledge together in an organized and coherent manner while building mathematics structurally from an axiomatic system. Now we shall look at several important contributions from Euclid's number theory.

An important theorem proved in the *Elements* was that there are an infinite number of prime numbers. Recall that a prime number is a positive integer that only has two unique factors. This is usually taken to be 1 and itself. Hence, the first prime numbers are 2, 3, 5, 7, 11, 13, 17, 19, 23, 29, 31, etc. Once every few years or so, we find a story in the news that the next prime number had been found. It is important to note that we shall never find a largest prime number, but rather the largest *known* prime number. This proof, one of the greatest of antiquity like the square root of 2 is irrational, is conducted in a similar manner. Before we conduct the proof, we first need to establish the fundamental theorem of arithmetic, which states that all numbers can be written as the product of prime numbers. For example, 6 can be written as $2 \times 3$ and 12 can be written as $2 \times 2 \times 3$. For a prime number such as 5, we simply write 5. To prove this notion, Euclid used a *reductio ad absurdum* proof by assuming the opposite: not all numbers can be written as the product of prime numbers. Let us state the smallest such number as $n$. That would mean that $n$, a positive composite integer, which is a positive integer that is not a prime number, could be written as $n = a \times b$, where both $a$ and $b$ are positive integers. Now, since $n$ is the smallest number that cannot be written as a product of prime numbers, $a$ and $b$ must be able to be written as the product of prime numbers. It is here we find our contradiction because this means that $n$ could be written as the product of prime numbers after all. Hence, we conclude that all numbers can be written as the product of prime numbers. Euclid continued his proof (not discussed in detail here) to show that any two representations as the product of prime numbers are, in fact, the same, which means that every number has a unique prime factorization.

For our proof that there is an infinite number of prime numbers, we first assume the opposite, that is, that there are a finite number of prime numbers, and proceed with a *reductio ad absurdum* proof. If there are a finite number of prime numbers, we can list them as follows: $p_1, p_2, p_3, \ldots, p_n$, where $p_n$ is the largest prime number. We can multiply all of the prime numbers together to yield: $p_1 \times p_2 \times p_3 \times \cdots \times p_n$. This new number is clearly composite because of all the prime factors. If we add 1 to this number, we yield: $p_1 \times p_2 \times p_3 \times \cdots \times p_n + 1$. This is where the proof becomes interesting. This new number is either a prime or composite number. If there are a finite number of prime numbers, we can expect this number to be composite. However, what factors other than 1 and itself could this number have? It clearly cannot have any of the prime numbers as factors: $p_1, p_2, p_3, \ldots, p_n$ because each of these prime numbers will be a factor of $p_1 \times p_2 \times p_3 \times \cdots \times p_n$. Since there are no prime factors of $p_1 \times p_2 \times p_3 \times \cdots \times p_n + 1$, we can conclude that this number is a new prime number. The contradiction is that we had already listed all prime numbers, and yet found a new prime number. Hence, there are an infinite number of prime numbers.

Euclid gave the result for the sum of geometric series, which was mentioned in Chapter 2 for the Eye of Horus. Recall that a geometric series is a series of numbers such that the ratio of any two consecutive terms gives us a constant ratio. We can represent a geometric series as follows: $a + ar + ar^2 + ar^3 + \cdots + ar^{n-1}$, where we can observe that the constant ratio is $r$. We can call this sum $S$, and then multiply $S$ by $r$ to get $rS = ar + ar^2 + ar^3 + \cdots + ar^{n-1} + ar^n$. Next we can subtract $S$ from $rS$ to yield $rS - S = a - ar^n$ because the middle term $ar$'s cancel out. If we factor $S$ on the left and $a$ on the right we yield $S(r - 1) = a(1 - r^n)$, and then divide both sides by $(r - 1)$ to yield $S = [a(1 - r^n)]/(r - 1)$ with the restriction that $r$ cannot equal 1. If the reader is wondering how we had $a/(1 - r)$ in Chapter 2 for the infinite geometric series, the answer is that we simply let $n$ get very large knowing that $r$ is a fraction and thus we lose the $r^n$ term (a fraction raised to a very large power is essentially zero for practical purposes). The geometric series can help contend with Zeno's paradoxes.

Another important contribution from the *Elements* is Euclid's algorithm for finding the greatest common divisor (GCD) of two numbers. Recall that the GCD of two numbers is the largest number that is a factor of both numbers. For small numbers such as 12 and 16, it is fairly easy to list the factors of 12 and 16 and find the number that works. For 12, we have factors 1, 2, 3, 4, 6, and 12, and for 16 we have factors 1, 2, 4, 8, and 16. Thus, 4 is our GCD. Finding the GCD for two numbers such as 42 and 120 would be more difficult. However, using Euclid's algorithm, we can subtract 42 from 120 to get 78. We can subtract 42 from 78 to get 36. Now, finding the GCD for 42 and 36 would be a much more reasonable expectation. We can factor 42 to yield 1, 2, 3, 6, 7, 14, 21, and 42. We can factor 36 to yield 1, 2, 3, 4, 6, 9, 12, 18, and 36. We can immediately see that 6 is our GCD. We find that, in general, any numbers, *a* and *b*, with *a* being smaller, have the same GCD as *a* and *a* − *b* do. We can continue this process until our numbers are small enough to make finding the GCD easier.

Before we leave Euclid, there are two interesting anecdotes that have been attributed to him that we will briefly explore. One says that King Ptolemy, who had commissioned the Lighthouse of Alexandria, one of the seven ancient wonders of the world, asked Euclid if there was an easier way to learn geometry, in which Euclid replied that there is no royal road to geometry. Here Euclid is making reference to the Persian Royal Road that was build for faster communication. The couriers on the Royal Road inspired the US Post Office motto that neither snow, nor rain, nor heat could prevent them from their delivery. Another story claims that one of Euclid's students asked him what benefits he would gain from learning Euclid's first theorem. Euclid responded angrily and said the student should receive a few cents since he feels he must gain something from what he learns. It is uncertain when Euclid died, but it was most likely around 265 BCE.

Our next mathematician, Archimedes, was referenced earlier as the greatest mathematician of antiquity. Archimedes was born in 287 BCE in Syracuse, which is in Sicily, Italy. He is known for his great mathematical accomplishments, as well as accomplishments in other areas including physics, engineering, and astronomy. Archimedes feared that others were stealing his work, therefore, he allegedly included a false theorem or two in his letters so that he would know if the mathematician he was corresponding with was stealing his work or not. Today, a picture of Archimedes is found on the prestigious Fields Medal, which is considered to be the Nobel Prize for mathematics. Archimedes discovered an early form of calculus integration to determine area. He divided a region into smaller pieces called "indivisibles" in order to find the area of a region that was difficult to determine otherwise. Some of the major mathematical contributions from Archimedes were finding the area of the circle, determining a good approximation for $\pi$ by inscribing a circle in polygons of larger and larger size, and finding the relationship between a sphere and a cylinder. He determined that $\pi$ was between

223/71 and 22/7. Much of the work of Archimedes involved circles; one of his works was titled *Measurement of a Circle*. We now look at the relationship between a sphere and a cylinder.

Archimedes found that if one were to place a sphere in a cylinder such that the sphere fits perfectly by touching the top, bottom, and sides of the cylinder, then the surface area and volume of the sphere would be two-thirds that of the cylinder. We can derive the surface area of a cylinder by imagining a soup can. The top and bottom circles both have $\pi r^2$ as their areas each, where $r$ is the radius. The surface area of the label is found by peeling it off and laying it down as a rectangle. The dimensions of the rectangle are the height of the can and the distance around the can, which is $2\pi r$. Thus, the area of the label would be $2\pi rh$, where $h$ is the height. Therefore, the surface area of the cylinder is $2\pi r^2 + 2\pi rh$. Since the sphere sits perfectly in the cylinder, we can conclude that the cylinder is $2r$ units in height, which gives us $2\pi r^2 + 2\pi r(2r) = 2\pi r^2 + 4\pi r^2 = 6\pi r^2$. Archimedes found that the sphere has two-thirds the surface area of the cylinder, which means the surface area of the sphere is $4\pi r^2$. It should be noted that this also appeared in Egypt in the *Moscow Mathematical Papyrus* well before Archimedes.

In terms of volume, we can consider that the volume of a cylinder is the area of either the top or bottom circle multiplied by the height, which gives us $\pi r^2 h$. Again, we know the height is $2r$ units, so we have $\pi r^2(2r) = 2\pi r^3$. Again, Archimedes found that the sphere has two-thirds the volume of the cylinder, which means that the volume of the sphere is $(4/3)\pi r^3$. The relationship between the sphere and cylinder was considered so important for Archimedes that a picture of a sphere inside a cylinder allegedly appeared on the tombstone of Archimedes.

Although Archimedes loved mathematics in a theoretical sense, he also appreciated the applications for engineering, which was rare in Greek thought. He is acknowledged as having created many inventions that benefited society. Two of particular interest were weapons of war to defend Syracuse from attack from the sea. Allegedly, Archimedes developed a claw that picked a ship out of the water in order to prevent it from attacking the shoreline. Another weapon he allegedly produced was the heat ray. Archimedes found that he could position large mirrors toward the sea to focus the rays of the Sun in order to burn wooden ships approaching from the sea. However, it is more likely that the mirrors would have only blinded a ship's crew instead of burning the actual ships.

Archimedes did not invent the lever, but he understood how it worked. He developed the idea for a fulcrum that he could easily use to lift objects that otherwise would be too heavy. He famously said that if given a level and a place to stand, he could move the Earth, and from a metaphorical perspective, he actually did accomplish this.

There are two final interesting anecdotes about Archimedes worth giving here. The first story involved Archimedes' use of "Eureka," which translates to "I have found it" in Greek. King Hiero II of Syracuse had given gold to a goldsmith to create a royal crown. Even though the crown weighed as much as the gold, Hiero was uncertain if the goldsmith mixed some silver into the crown. Hiero asked Archimedes if he could determine the gold content of the crown without destroying the crown. Archimedes discovered how to do this while getting into a bathtub. Archimedes realized that his body displaced a certain amount of water and that since gold is heavier than silver, more silver would have been needed than the equivalent amount of gold if the crown were impure. Archimedes would simply need to see if the crown displaced as much water as a piece of pure gold equivalent in weight, which today we call Archimedes' principle. Allegedly, he was so excited by his discovery that he ran out of the bath naked down the streets of Syracuse yelling, "Eureka! Eureka!"

The second story involves the death of Archimedes in 212 BCE. During the Siege of Syracuse, a Roman soldier was sent to get Archimedes, but was ordered not to harm him. The soldier found Archimedes on the beach working out his mathematics in the sand. Archimedes was so involved in his work that he failed to realize the city had been invaded by the Romans. Allegedly, the soldier commanded Archimedes to come with him, but Archimedes told him to wait until he was finished. The soldier became infuriated and killed Archimedes right on the spot on the beach. Legend has it that the last words of Archimedes were to tell the Roman soldier not to disturb his circles, in reference to the work he was doing with the circles on the beach.

The next mathematician we shall examine is Eratosthenes, who was born in 276 BCE in Cyrene, which is in modern-day Libya. Eratosthenes' greatest contribution was to geography, a term coined by him. He was the first person to prove the Earth was roughly spherical and to calculate the circumference of the world. He was the first to create an accurate map of the known world. Eratosthenes is also known for the Sieve of Eratosthenes, which was a method of finding prime numbers. Eratosthenes started by crossing out all multiples of 2, except for 2 itself, which is a prime number so this number gets circled. Next, he circled 3, another prime number, and crossed out all multiples of 3. Next he circled 5, which is a prime number because it was not yet crossed out. Eratosthenes would have already crossed out 4, since it is a multiple of 2. Next, he crossed out all multiples of 5. This meant that the next number not crossed out was a prime number. Eratosthenes was also known as Beta because even though he excelled in many fields, he was never considered the best in any area. He later became blind and died around 195 or 194 BCE, possibly by starving himself to death.

It is worth noting here that there is a common misconception that many people believed the world was flat during the time of Christopher Columbus in the 15th century CE. Eratosthenes' proof that the Earth was roughly spherical, however, tells us that people knew the world was round since ancient times. The main issue for Columbus was not so much a flat or round Earth, but the existence of the Americas. Columbus believed that sailing across the Atlantic Ocean would bring him to Asia. Indeed, the debate that would occur sometime after Columbus was the view of a geocentric or heliocentric solar system. We shall address this conversation later.

Our next mathematician is Apollonius, who was born in 262 BCE in Perga, which today is in modern-day Turkey. Apollonius is known as "The Great Geometer" for his contributions to geometry and, in particular, to conic sections. Conic sections are curves that are made by a plane intersecting the cone. Apollonius outlined the conic sections in his work, *Conics*, which was written in eight books (only seven have survived). The conic sections include the parabola, circle, ellipse, and hyperbola. Additionally, Apollonius improved on the approximation of $\pi$ made by Archimedes. He died around 190 BCE.

The Romans finally conquered Greece in 146 BCE at the Battle of Corinth. Corinth, which had been very wealthy with a high standard of living, was completely destroyed by the Romans. This marked the end of the Hellenistic Period. However, as previously stated, the Romans accepted and adopted much of the Greek culture and hence the period following has been named Roman Greece. For our purposes in this book, the spirit of the Hellenistic Period lived on and there would be further mathematical developments in the Greek tradition.

It should be acknowledged that all of the major mathematical developments from this time onward were in Greek tradition and that the Roman culture itself did not develop any type of innovative mathematics. Mathematics was an important part of Roman education because Rome adopted the quadrivium of arithmetic, geometry, astronomy, and music. Although a base 10 system, the Roman numeral system (Table 6.1), very likely familiar to the reader, was not very innovative for its time and was not a true positional system. It was

**TABLE 6.1   Roman Numerals**

| Number | Symbol |
| --- | --- |
| 1 | I |
| 5 | V |
| 10 | X |
| 50 | L |
| 100 | C |
| 500 | D |
| 1000 | M |

additive, but not multiplicative as is the system we use today. The Roman numeral system was additive in that the symbol for 2 was the combination of two ones: II. Six was written as 5 + 1: VI. Sometimes subtraction was necessary for representation such as 9, equivalent to 10 − 1, and the 1 is on the left of 10 to indicate this: IX. Computations with Roman numerals were quite cumbersome. It is easy to imagine the difficulties that lie in multiplying three digit numbers using Roman numerals. Fractions in Roman numerals were based around the idea of twelfths since 12 has a large number of factors.

The first mathematician we shall address after the fall of Greece is Claudius Ptolemy, who was born in 90 CE in Egypt, but was a Roman citizen and wrote in the Greek tradition. He is known to be the "Father of Trigonometry" due to his work with triangles and development of work that led to modern trigonometric formulas. He is best known for his books the *Almagest* and *Geographia*. The *Almagest*, or originally known as the *Mathematical Collection*, was an important book of Greek astronomy, and was highly influential for its geocentric model (i.e., the heavenly bodies revolve around the Earth) inherited from the Babylonians. It was not until the 16th century that Copernicus would promote the heliocentric model, which indicates that the heavenly bodies, including the Earth, revolve around the Sun. *Geographia* was a study in geography and influential in map making during the Roman Empire. It introduced the concept of latitude and longitude on the Earth. Ptolemy died around 165 CE.

Diophantus was born in Alexandria around 200 CE, and is unique for his work in algebra. Much of the Greek mathematicians were primarily concerned with geometry, but due to Diophantus' work in algebra, he is often called the "Father of Algebra." However, Diophantus collected much of what was already known in algebra, primarily from the work of the Babylonians. His famous work was *Arithmetica*, which is a collection of algebraic problems that involved the solving of equations. In *Arithmetica*, we are introduced to Diophantine equations, which are algebraic equations where the variables must be integers. Diophantine equations have more variables than equations, so it is often possible to find many integer solutions.

A major contribution to algebra was his early use of algebraic symbols, and Diophantus worked with quadratic equations in three forms:

**1.** $ax^2 + bx = c$,
**2.** $ax^2 = bx + c$,
**3.** $ax^2 + c = bx$.

A classic algebra problem allows the reader to calculate how old Diophantus was when he died:

Here lies Diophantus, the wonder behold. Through art algebraic, the stone tells how old: God gave him his boyhood one-sixth of his life. One twelfth more as youth while whiskers grew rife. And then yet one-seventh ere marriage begun. In five years there came a bouncing new son. Alas, the dear child of master and sage after attaining half the measure of his father's life chill fate took him. After consoling his fate by the science of numbers for four years, he ended his life.

Algebraically, we can represent this as follows if we allow $x$ to be the length of his life in years: $(1/6)x + (1/12)x + (1/7)x + 5 + (1/2)x + 4 = x$. Solving for $x$ yields $x = 84$. Hence, Diophantus lived for 84 years.

Hypatia was born in Alexandria around 370 CE, and she is considered the first known woman to make significant contributions to mathematics. Her father, Theon, was a mathematician and philosopher who probably provided her with an education. Hypatia may not have produced her own original work, but she wrote many commentaries on works such as Euclid's *Elements*, Archimedes' *Measurement of a Circle*, Apollonius' *Conics*, Ptolemy's *Almagest*, and Diophantus' *Arithmetica*.

Hypatia was not a Christian, but she had some Christian students. In 415 CE, a group of angry Christian monks, likely led by Peter the Reader, attacked and killed Hypatia. Some historians contend that this was because of her friendship with Orestes, the Roman governor of Egypt, who happened to be a Christian. There was a feud between Orestes and Cyril, bishop of Alexandria. However, some historians believe the Christian monks were threatened by Hypatia's scholarship. There has been an ongoing debate regarding the true motivation, but it is generally accepted that Hypatia was killed at least partially for being a non-Christian woman of advanced scholarship. There is also some debate as to how she was killed, but generally it is accepted that she was stripped of her clothing in public before being dragged through the streets and brutalized. Eventually, her remains were burned.

Some historians place the death of Hypatia as the decline of Classical Greek thought, but not the end. It certainly contributed to the decline and end of Greek mathematics at Alexandria. The Library of Alexandria was burned by Muslim conquerors in 641 CE. It is possible that the caliph believed if the books in the library agreed with Islam, then they were redundant to the Qur'an. If they disagreed, then they should be burned for their contradiction to the Qur'an. The Greek Period continued until about the time of the fall of the Roman Empire to the barbarian invasions in 476 CE, and it could be argued that the Greek Period continued even until 529 CE when Byzantine Emperor Justinian I finally closed Plato's Academy. The Byzantine Empire, also known as the Eastern Roman Empire, was centered around Constantinople in modern-day Turkey. It split from the Roman Empire in 395 CE, but continued to flourish, well after the fall of Rome, until 1453 CE, when Constantinople fell to the Muslim Ottoman Turks.

Shortly after the fall of Rome in 476 CE, a philosopher was born near Rome named Boethius who studied Greek and was familiar with the classical Greek works. Although his mathematical works were not of very high quality, they continued to influence Europe for many centuries due to a lack of better alternatives. Boethius was executed for treason by Justinian I in Constantinople around 524 CE.

As previously stated, the decline of Greek thought can be attributed to competing ideas of Christianity in the first several centuries of the Common Era, culminating in Emperor Constantine's adoption of Christianity in 312 CE and Christianity becoming the official state religion of the Roman Empire in 380 CE, which furthered the decline of Greek thought. Christianity was generally not concerned with scientific thought because the Bible provided people with all of the knowledge they needed. People did not need to use reason to find the truth because God spoke to the people through the Bible. Furthermore, rewards were in the afterlife, not here on Earth. This was in direct conflict with the way the Greeks thought, and we generally would not see Christian acceptance of Greek philosophy until Thomas Aquinas embraced the reason and philosophy of Aristotle in the 13th century CE. Rome proper did not produce any significant mathematics outside the Greek-dominated territories. It is possible that this can be attributed to the attitude of the ruling class, who relied on the laboring class to do the work of the empire and felt little need to develop intellectual advancement similar to the Greeks. After the fall of Rome, there was very little scientific development in the Western world until nearly the time of the Renaissance in the 14th century CE. This period has been called the Dark Ages, although this is generally not an accepted term by most historians.

In the next part of the book, we look to Chinese, Indian, and Islamic thought spanning ancient time to the Middle Ages in which advancements in mathematics continued and later were brought to Europe by the Muslims before the Renaissance. In the next several chapters, we shall explore the contributions of these cultures to mathematics, and we shall also explore mathematics in Pre-Columbian America.

# MATHEMATICS IN ASIA AND PRE-COLUMBIAN AMERICA

# CHAPTER 7

# CHINESE MATHEMATICS: ANCIENT TIMES TO THE MIDDLE AGES

Up until now, we generally followed a linear path of development starting with the Egyptians and Babylonians and then onto the Greeks. In East Asia, Chinese mathematics had reached great developments from ancient China throughout the Middle Ages. The Chinese, like the Egyptians and Babylonians, were concerned with the practical use of mathematics, such as in astronomy for calendar making, in engineering for practical purposes, and informing civil service functions in a centralized bureaucracy. Calculation and problem solving was an important part of Chinese mathematics. While the Chinese did not employ a proof system, as did the Greeks, they did, in fact, develop very sophisticated mathematics.

Chinese civilization first developed along the Yellow River. There is some evidence that the first Chinese dynasty was the Xia Dynasty that began about 4000 years ago and flourished for about 500 years. The Xia Dynasty was followed by the Shang Dynasty that started around 3500 years ago, and it is during the Shang Dynasty that the first written records are found through use of markings on animal

*The Development of Mathematics throughout the Centuries: A Brief History in a Cultural Context*, First Edition. Brian R. Evans.
© 2014 John Wiley & Sons, Inc. Published 2014 by John Wiley & Sons, Inc.

bones, including evidence of mathematics that described an additive and multiplicative base 10 system of numeration, which will be addressed at the end of this chapter.

The Zhou Dynasty spanned from about 1000 to 250 BCE and was a period of great cultural advancement. It marked the beginning of Chinese philosophy. It was during the Zhou Dynasty that Confucius, also known as Kong Qui, had lived. Confucius was a Chinese philosopher who was born in 551 BCE in Qufu in eastern China. His philosophy emphasized learning and ethical behavior, and he was an early developer of the golden rule. Confucius died in 479 BCE and his teachings evolved into the philosophy of Confucianism. Living around the same time was Sun Tzu, author of the *Art of War*, and Lao-Tse, founder of Taoism. It was during this time that much of the Great Wall of China was constructed. A calendar system that was both lunar and solar was developed during the Zhou Dynasty in about 500 BCE.

During the Zhou Dynasty, students studied the Six Arts, which would be analogous to the quadrivium of Greece and Rome, and it included calligraphy, music, religion, archery, the chariot, and mathematics. Studying the Six Arts made a person well rounded, and expertise in each made a person someone who would be called a "Renaissance Man" in the West.

After warring factions disintegrated the Zhou Dynasty, the short-lived Qin (or Chin) Dynasty consolidated ancient China, completed more work on the Great Wall, and built the Terracotta Warrior statues in Xi'an. It was during the Qin Dynasty that the first Emperor of China, Qin Shi Huang, rose to power. The Western name for China comes from the Qin Dynasty, even though the Chinese call the country Zhongguo, originating from the Zhou Dynasty, which means "middle kingdom."

Qin Shi Huang was accused by the following dynasty of using brutal methods to maintain power. It is believed that Qin Shi Huang required the burning of all books to keep the people under control, but it is not certain if this actually occurred. If the burning of the books did take place, some works surely survived to be complied during the next dynasty. The centralized bureaucracy of the Qin Dynasty provided a purpose for the development and study of mathematics over the next several centuries. It was during this dynasty that weights, measures, currency, and written language were standardized. Not long after the short-lived Qin Dynasty, a very significant dynasty would rule China between 200 BCE and 200 CE.

The Han Dynasty saw trade between China and the West progress through the use of the Silk Road. There were great scientific advancements, including the use of iron and steel; innovations in agriculture, astronomy, and engineering; and the invention of paper. The Han Dynasty witnessed much advancement in mathematics including the compilation of mathematical books and early use of the abacus. It is during this period that bamboo counting rods were used to indicate place value with an empty space left to indicate zero, an early innovation in the historical development of the concept of zero, which is quite sophisticated. The Chinese developed methods for solving quadratic equations during the Han Dynasty. One of the early works that remains is the *Zhoubi Suanjing*, or *Shadow Manual*, a text of Chinese astronomy written between 100 BCE and 100 CE. The *Shadow Manual* contains the Chinese version of the Pythagorean theorem, called the Gougu rule, but it is debatable whether or not a proof is actually provided. The Gougu rule was probably known much earlier than the *Shadow Manual* was written, and may even predate Pythagoras.

The most important and well-known Chinese mathematics book is the *Jiuzhang Suanshu*, or *Nine Chapters on the Mathematical Art*, which was compiled from works of many authors and developed over 1000 years. It was put in a more finalized form during the first century of the Common Era. It contained 246 mathematics problems in nine chapters and for Chinese mathematics is analogous to what the *Elements* was for Western mathematics. The *Nine Chapters* was concerned with practical issues in surveying, agriculture, engineering, finance, and civil affairs. It was the first book in the world to address the concept of negative numbers, which came about around 200 BCE. It portrayed Gaussian elimination for solving a system of linear equations possibly over 2000 years before it would be developed in Europe by Gauss, who may have been the greatest mathematician of all time and is discussed in a later chapter. It is notable that like the ancient works mentioned in earlier chapters, the *Nine Chapters* contains Pythagorean triples. This remarkable collection in Chinese mathematics was highly influential for 1500 years, and it is generally accepted that until the completion of the *Nine Chapters*, around the first century of the Common Era, mathematical development in the Western world and China were independent.

The first recorded magic square (Figure 7.1) in China appeared in the *Nine Chapters*, but probably dates back to around 650 BCE. The magic square, known as the Lo Shu square, is a $3 \times 3$ square in which the rows, columns, and diagonals sum to 15 using the integers 1 to 9 only once for each space. Legend has it that a turtle, used for offerings to the gods, emerged from a river with the magic square pattern on the back of the shell.

Born around 220 CE in Wei, which is located in northern China, after the end of the Han Dynasty, Liu Hui was one of the greatest mathematicians of ancient China. At the end of the Han Dynasty, China was divided and remained that way for nearly 400 years until

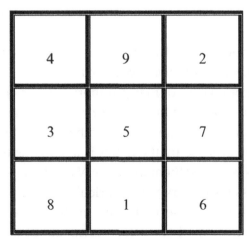

**FIGURE 7.1**   Chinese Magic Square

the Sui Dynasty. Liu Hui is famous for his commentary and solutions to problems from the *Nine Chapters*. While the *Nine Chapters* were primarily concerned with mathematics problems without necessarily providing justification, Liu Hui worked to justify claims in the *Nine Chapters*. He provided a justification for the Pythagorean theorem in addition to calculating a fairly accurate approximation of $\pi$ similar to Archimedes but with more accuracy, which was an improvement on the value of 3 used in the *Nine Chapters*. Liu Hui's addendum to the *Nine Chapters* is called the *Haidao Suanjing*, or *Sea Island Mathematical Manual*, which included nine additional problems to the *Nine Chapters*. The book is named after the first problem that involved the calculation of distance in relation to an island. Liu Hui died around 280 CE.

The next Chinese mathematician we shall discuss is Zu Chongzhi, who was born in 429 CE in Jiankang, which is in northern China. He came from a family of astronomers who specialized in calendar making, and Zu Chongzhi was able to produce an accurate approximation for the length of the year in days that is very close to the value we use today. Like Lui Hiu, Zu Chongzhi was one of the greatest mathematicians of ancient China. He learned the mathematics of Lui Hiu and used his method of calculating $\pi$ to find an even more accurate approximation than Lui Hiu found. Zu Chongzhi discovered $\pi$ to be about 3.1415926, and this was regarded as the most accurate value for the next 900 years. He also found the best rational number approximation for $\pi$ at that time, which is $355/113$. He died around 500 CE.

In 589 CE, the Sui Dynasty reunited the country, but this dynasty did not last very long. By early 7th century, the Tang Dynasty was ruling China and the country experienced great prosperity. It was during this dynasty when the Chinese invented woodblock printing, well before Gutenberg's printing press in Europe. It was during this time that the *Ten Classics* were compiled, which consisted of the 10 great Chinese mathematical books, including the *Nine Chapters*. The *Ten Classics* were used for preparation for the civil service examinations.

The Tang Dynasty ended in 960 CE and was succeeded by the Song Dynasty, which was a dynasty of great scientific progress with the invention of gunpowder and the magnetic compass. Jai Xian was born around 1010 CE and developed Pascal's triangle about 600 years before Pascal would develop his triangle in Europe. Yang Hui, who was born in 1238 CE and died in 1298 CE, referenced the work of Jai Xian, and the Chinese version of Pascal's triangle is called, unfortunately for Jia Xian, Yang Hui's triangle. Jia Xian is also remembered for his solutions of square and cube roots. He died around 1070 CE. However, it was at the end of the Song Dynasty, which ended in 1279 CE, that Chinese mathematics entered a Golden Age in the 13th century CE.

Born in 1202 CE in Ziyang, Qin Jiushao is considered to be one of China's most important mathematicians. However, he did not devote his entire life to mathematics; he also served as a politician. Unfortunately, as a politician, he was known for corruption, brutality, and for poisoning his enemies. His famous mathematical work is *Shushu Jiuzhang*, or *Mathematical Treatise in Nine Sections*, which was published in 1247 CE. As other works in the Chinese mathematical tradition, this work was concerned with practical mathematics such as solving equations, finance, and construction. The work expanded on the Chinese remainder theorem, a concept in number theory that had existed in China for nearly 1000 years. Qin Jiushao also worked with quadratic and cubic equations. He is notable for his use of a zero symbol as a placeholder. Prior to his work, only a blank space was used. It should be noted, however, that the mathematics of Qin Jiushao would not be seen in Europe until centuries later. Qin Jiushao died in 1261 CE.

Li Zhi, also known as Li Yi, was born in 1192 CE and died in 1279 CE around the end of the Song Dynasty and the beginning of the Yuan Dynasty, which began as a result of the Mongol invasion of China in 1271 CE, 64 years after the death of Genghis Khan in 1227 CE. Li Zhi is best known for his work in solving polynomial equations up to the sixth degree. Another mathematician in the Golden Age is Zhu Shijie, who was born around 1260 CE and died around 1320 CE. He is remembered for his development of Chinese algebra.

Chinese mathematics continued throughout the next several centuries, but the Golden Age of Chinese mathematics was very clearly centered in the 13th century CE. By the 14th century CE, Chinese mathematics went somewhat into decline. China increased contact with the West and was introduced to other mathematical ideas. Although Western methods were respected, Chinese mathematicians retrained their mathematical traditions. Beginning in the mid-20th century CE, Chinese mathematicians became quite active in the global mathematics community and have taken a leading role in 21st-century mathematics.

Let us now move on to the specifics of the Chinese number system. In our time, there has been a debate in mathematics education in the United States on the best way to improve American students' test scores in mathematics and science in relation to the high achieving students in parts of East Asia. Interestingly, one reason given for the difference in achievement may have to do with the way numbers are pronounced. For example, in English and many European languages the number 11 has its own unique name. In English, this is "eleven" and in Spanish, this is "once," which is pronounced "un-say." However, in Chinese, 11 is said as "ten one" and 12 is said as "ten two." Twenty is said as "two ten" and 30 is said as "three ten." Furthermore, in Chinese, each number has a very short monosyllabic word as its name, unlike a number such as "seven" in English. Researchers believe that this simplified system makes computations easier and may free the mind for more advanced mathematics. Now that we have discussed how Chinese numbers are named, let us take a look at one of the ancient Chinese methods of writing numbers and its numeration system.

The Chinese developed a numeration system that was both additive and multiplicative, and it is possible that this number system is 3500 years old. The aspect of this system that distinguished itself from the systems that developed in the West is the multiplicative nature of this system. For example, today we write the number 52 using a 5 and a 2 in the correct position, meaning we have five 10s and two 1s. The Egyptians would not write 52 using the multiplier 5, but rather use their symbol for 10 by writing it out five times followed by a 1 written two times. The innovative Chinese system was more multiplicative than positional in that today we do not need to write the 10 next to the 5 in 50, but rather we know based on the position of the 5 that we have five 10s. In the Chinese system, we would have the symbol for 5 positioned next to the symbol for 10. It is important to note that the Chinese system was written vertically. The number 52 would appear as follows:

5
10
2

Notice that the 5 written over the 10 means five 10s, and the 2 indicates we should add two more. However, notice that there was no 1 symbol used to indicate that there were only two 1s, which was deemed unnecessary. Another interesting aspect of the Chinese system is that if we have a number such as 314, we would have only one 10. We would not need to indicate that there was only one 10; rather, just using the symbol for 10 means the 1 is implied:

3
100
10
4

However, it is interesting to note that the number 324 would be written as follows, which uses the multiplier 2 with the 10:

3
100
2
10
4

The Chinese had a special symbol for zero that was used to indicate that a power of 10 was missing. It is important to note that this symbol for zero was not an actual number. This symbol was used only when the power of 10 was missing and was not used as a true positional zero. If the 1s or the 1s and 10s were missing, such as 340 or 300, the zero symbol was not needed. For 340 we would have the following:

3
100
4
10

For 300, we would have the following:

3
100

However, the number 305 would be represented as follows:

3
100
0
5

A number such as 6007 would be represented as follows. It is important to note that only one zero symbol is used to represent zero 100s and zero 10s.

6
1000
0
7

Let us look at the actual symbols used by the Chinese followed by the Chinese representations of the numbers that were already mentioned (Table 7.1).

**TABLE 7.1   Chinese Numerals**

| Number | Symbol | Number | Symbol |
|--------|--------|--------|--------|
| 0 | 零 | 7 | 七 |
| 1 | 一 | 8 | 八 |
| 2 | 二 | 9 | 九 |
| 3 | 三 | 10 | 十 |
| 4 | 四 | 100 | 百 |
| 5 | 五 | 1000 | 千 |
| 6 | 六 | 10,000 | 万 |

Chinese representation for 52:

五
十
二

Chinese representation for 314:

三
百
十
四

Chinese representation for 324:

Chinese representation for 340:

Chinese representation for 300:

Chinese representation for 305:

Chinese representation for 6007:

A final interesting note about Chinese mathematics involves the tangram set (Figure 7.2). The tangram set is a set of seven shapes: two large triangles, one medium triangle, two small triangles, a square, and a parallelogram. The legend says that a servant of an ancient Chinese emperor was carrying an expensive ceramic square tile for the emperor. The servant accidentally dropped the tile on the floor and it shattered in seven pieces. The servant desperately tried to put the pieces back into a square, but while creating many other interesting shapes could not create the square again. Today, a student in a

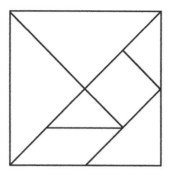

**FIGURE 7.2**    Tangram

mathematics class uses the tangram set to make various shapes and finally recreate the square.

In the next chapter, we shall explore the mathematical developments in India from ancient times to the Middle Ages. We shall see other similar great developments in mathematics in the southern subcontinent region of Asia.

# CHAPTER 8

# INDIAN MATHEMATICS: ANCIENT TIMES TO THE MIDDLE AGES

As with Chinese mathematics, Indian mathematics developed independently of the mathematics that was advancing in the eastern Mediterranean and the Middle East. Since India and China are separated by the Himalayan mountain range, the highest mountain range on Earth, it is very possible that the mathematics of India and China developed independently just as their cultures and societies developed independently. Like the mathematics of China, Egypt, and Babylon, Indian mathematics was developed for practical purposes, but with more attention to religious affairs, such as the geometric design of religious altars. Much of early Indian learning was transmitted orally through memorization before the introduction of a writing tradition.

Indian civilization developed along the Indus River on the Indian subcontinent around 5000 years ago. There is evidence that this civilization created a systemic means of weights and measures. Around 3500 years ago, the Vedic culture developed, sharing its name with the religious texts composed during this period, the *Vedas*. The *Vedas*, comprised of works written in Sanskrit from oral tradition between

*The Development of Mathematics throughout the Centuries: A Brief History in a Cultural Context*, First Edition. Brian R. Evans.
© 2014 John Wiley & Sons, Inc. Published 2014 by John Wiley & Sons, Inc.

1000 and 500 BCE, are a religious collection, and it is from Vedic religion that Hinduism evolved. Around the same time period, we see one of the earliest works in Indian mathematics to emerge, called the *Sulbasutras*, which were books describing the geometry necessary for the building of religious altars. However, it is likely that the geometry of the *Sulbasutras* was known well before the work was composed. One of the authors of the *Sulbasutras* was named Baudhayana, and it is in Baudhayana's work that we see Pythagorean triples and the earliest recorded statement of the Pythagorean theorem. Baudhayana also approximated $\pi$ and provided a method for finding the square root of 2.

The religion Jainism developed over the period spanning 1000 to 500 BCE. Mahavira, who lived between 599 and 527 BCE, established the core beliefs of Jainism. The core values emphasized nonviolence toward all beings, strict vegetarian diets, and nonattachment to material possessions. Jain mathematics brought Indian mathematics out of purely religious purposes, and expanded its use in arithmetic, geometry, algebra, and perhaps, most interestingly, combinatorics. The night sky, like many civilizations, influenced Jain developments in astronomy and mathematics. Jain mathematics used the square root of 10 as an approximation for $\pi$.

Siddhartha Gautama, more commonly known as the Buddha, which means the "enlightened one," lived in India in the 6th and 5th centuries BCE. He taught that suffering was part of the existence and that people should strive to eliminate desire. The Buddha is known for his Four Noble Truths regarding the origin of suffering and the elimination of it. The fourth of the Noble Truths is the Eightfold Path, involving right thoughts and actions, which leads to the elimination of suffering. He taught the importance of moderation and ethical behavior.

Sometime during the last several centuries before the Common Era, the *Bhagavad Gita*, which is a very significant Hindu text, was written. Around this time, an early Indian base 10 system developed. This system had unique symbols for the numbers 1 to 10, and also for 20 to 100. For numbers greater than 100, a system similar to the Chinese multiplicative system was employed. The numerals from this period are called the Brahmi numerals, and they would later evolve into today's Hindu–Arabic numeral system. It should be noted that while the symbols evolved into Hindu–Arabic numerals, the system itself changed quite a bit due to the birth of a positional number system.

Around 500 BCE, the Persian Empire invaded northwestern India, and as discussed earlier in Chapter 5, Alexander the Great made his conquests as far east as northwestern India around 330 BCE. Both the Persian and Greek cultures had an impact on Indian culture. After the death of Alexander the Great, the Maurya Empire controlled India between 322 and 185 BCE. The Maurya Empire continued to trade and have relationships with the West, which meant there was ongoing contact and influence between the cultures.

In the centuries that followed, India remained fragmented but was reunited by the Gupta Empire, which ruled between the 4th and 6th centuries CE. This ushered in the Classical or Golden Age of India, during which great accomplishments in literature, art, philosophy, science, and mathematics were made. The great prosperity and peace during this time allowed for significant accomplishments to be attained during this period. It has been claimed that during the Gupta Empire, the world's first truly positional base 10 system began to develop along with the concept of zero as a placeholder. However, other evidence points to this development sometime during the several centuries before the Common Era. India's Classical Age was also the beginning of India's Classical Age of mathematics as well. However, the Classical Age of mathematics lasted for centuries more until about the end of the 12th century CE.

Around 400 CE, a work with unknown authorship developed, called the *Surya Siddhanta*, which formed the basis for modern trigonometry. It was primarily a book in astronomy, but introduced the concepts of sine and cosine. It is from this work in trigonometry that we get the words for sine and cosine, which are "jya" and "kojya," respectively. "Jya" was mistranslated into Arabic to be "jaib," which means "fold." In Latin, this would be "sinus," hence our use of the word "sine." Sometime after the *Surya Siddhanta*, another work,

*Lokavibhaga*, probably written in 458 CE, was the first text to use a base 10 positional system and to make reference to zero as a placeholder.

The Hindu–Arabic numeral system developed over centuries into the system we use today, which consists of nine digits and zero. The truly remarkable aspect of the system is that it is positional. Recall in Chapter 2 that the concept of a positional system was presented. For example, in the Hindu–Arabic system, we do not have to write 823 as 100 eight times, 10 two times, and 1 three times. Further, we do not have to write 823 as $8 \times 100 + 2 \times 10 + 3 \times 1$, but rather we know the position of the individual digits represents 100s, 10s, and 1s. For example, 823 is quite different from 382. In the first number, we know we have eight 100s, two 10s, and three 1s. In the latter number, we have three 100s, eight 10s, and two 1s. For us, it is the position of the digits that make the difference. Zero has a critical role in the positional system. For example, writing 803 means we have eight 100s, no 10s, and three 1s.

There is no doubt that the first development of this system occurred in India. This system is called "Hindu–Arabic" because it was essentially developed in India and then adopted and refined by Islamic mathematicians, who added the use of decimal fractions to this system. It is the Islamic mathematicians who brought the system to Europe, and this will be addressed further in the next chapter.

During the Gupta Empire, India's first great mathematician of the Classical Age, Aryabhata, was born in 476 CE in what is now Patna in eastern India. He is known for his early use of a space for a placeholder that would eventually lead to zero as a placeholder. He also developed a good approximation for $\pi$, which was $62{,}832/20{,}000$. Aryabhata also contributed to algebra, trigonometry, astronomy, and differential equations. He was an early mathematician to work with the arithmetic series. Around the time of Aryabhata, the relationship between sine and cosine, $\sin^2 x + \cos^2 x = 1$, was discovered by Varahamihira, a mathematician born around 505 CE and died 587 CE, who was also known for innovation with Pascal's triangle. Aryabhata died in 550 CE.

After the demise of the Gupta Empire, Indian mathematics continued to flourish. The next great Indian mathematician from the Classical Age of mathematics is Brahmagupta, who was born in 598 CE in northern India. Brahmagupta worked in Ujjain in central India, which was the center for mathematics and astronomy in India beginning about the 6th century CE. Brahmagupta's most famous work was *Brahmasphutasiddhanta,* and he was among the first mathematicians, along with Bhaskara I during the same time period, to use zero as a number and to introduce the concept of additive identity: $a + 0 = 1$. This means Brahmagupta used zero as both a number and a placeholder. Another major contribution from Brahmagupta was the beginnings of the quadratic formula, which would be expanded on by Perisian mathematicians al-Khwarizmi in the 9th century CE and Spanish Jewish mathematician Abraham bar Hiyya in the 12th century CE. The quadratic formula would be fully developed by Simon Stevin from Belgium in 1594 CE. Although Brahmagupta's quadratic formula could not generalize as it can today, it was the first appearance of a mostly generalized formula and involved the use of both negative and positive roots. The modern quadratic formula solves a quadratic equation of the form $ax^2 + bx + c = 0$, in which we can find the roots using the Babylonian method of completing the square. If we divide each term by $a$ on both sides of the equals sign, we yield

$$x^2 + \frac{b}{a}x + \frac{c}{a} = 0$$

Now that the coefficient in front of the first term is 1, we can complete the square. We need to divide the middle term by 2, square the result, and add this to both sides of the equation. We shall also remove $c/a$ from both sides. This gives us

$$x^2 + \frac{b}{a}x + \frac{b^2}{4a^2} = -\frac{c}{a} + \frac{b^2}{4a^2}$$

On the left side, we can factor and yield $(x + b/2a)^2$ and on the right we can find a common denominator and yield $(-4ac + b^2)/4a^2$. Taking the square root of both sides gives us $x + b/2a = \pm\sqrt{b^2 - 4ac}/2a$ and subtracting $b/2a$ from both sides gives $x = \left(-b \pm \sqrt{b^2 - 4ac}\right)/2a$.

Brahmagupta also provided us with a formula for finding the area of a rectangle using the sides of the rectangle. This can be recognized as a more general case of Heron's formula used to find the area of a triangle. Heron was a Greek mathematician in the 1st century of the Common Era. Brahmagupta died around 670 CE.

In the 8th century CE, Muslim invaders occupied modern-day Pakistan. This began the Muslim conquests throughout India over the next several centuries, including Delhi in the 12th century CE. However, despite the many changes in Indian society, mathematics carried on through its own Classical Age well past the time the Gupta Empire had expired. In the 9th century CE, Mahavira, born around 800 CE and died around 870 CE, was considered to be the last of the great Jain mathematicians. He wrote *Ganita Sara Samgraha*, which was comprehensive of 9th-century CE Indian mathematics and was meant to be an update to Brahmagupta's work. It was during the 9th century CE that the world first saw the symbol "0" used for zero. Evidence can be found of the symbol for zero on the walls of the Chaturbhuja Temple, which is a temple dedicated to Vishnu in central India. However, zero as an actual number probably originated in the middle of the 7th century in India.

Born in 1114 CE in southern India, Bhaskara II is considered to be the last mathematician of the Classical Age of mathematics for India and is regarded as India's greatest mathematician during the Middle Ages. He is likely considered to be the greatest Indian mathematician of all time. Like Brahmagupta, Bhaskara II worked in Ujjain. He made contributions to arithmetic, algebra, geometry, trigonometry, astronomy, and calculus. It could be argued that Bhaskara II developed differential calculus 500 years before Leibniz and Newton found differential and integral calculus in Europe, both of whom are discussed in a later chapter. While it can be argued that calculus was first developed in India, it is widely accepted that Leibniz and Newton unified calculus more comprehensively without direct Indian influence. Bhaskara II is credited to be the first to realize that an integer divided by zero approaches infinity. To understand this, consider that $1/10 = 0.1$, $1/100 = 0.01$, $1/1000 = 0.001$, and so on. This means that as the denominator grows larger, or approaches infinity, the overall result approaches zero. Conversely, we could have $1/(1/10) = 10$, $1/(1/100) = 100$, $1/(1/1000) = 1000$, and so on. This means that as the denominator gets smaller, or we can say approaches zero, the overall result gets larger. Hence, we can conclude that as the denominator approaches zero, the overall result approaches infinity, and as the denominator approaches infinity, the overall results approaches zero. It is important to stress that the denominator *approaches* zero, but does not equal zero. Otherwise, the results would be undefined. We can examine this in the following way. Let us say we have $8/4 = 2$. Another way of stating this is that $4 \times 2 = 8$. What if we have $8/0$? We can imagine that $8/0$ is some unknown quantity $x$. Thus, we have $8/0 = x$. Another way of stating this is 0 times $x$ equals 8, or $0x = 8$. What possible value could $x$ be? Any value substituted for $x$ would always yield zero, and never yield 8. This establishes that any nonzero number divided by zero has no solution, but what about $0/0$? We might get $0x = 0$, and of course a possible value for $x$ could be zero. However, any value would still make the statement true. This means that there is no unique value for $x$, which means that we can conclude again that there is no solution. This seemed to bother Bhaskara II, who would have preferred to divide by zero.

Bhaskara II provided a unique and interesting proof for the Pythagorean theorem that involved only one word, "Behold!" (Figure 8.1). Bhaskara II felt no need to give more instruction because he believed that the proof would be obvious once the picture is seen. The proof involves all four triangles being congruent and right. If the reader labels the sides of the big square as $c$, the bases of the right triangles $a$, and the height of each right triangle $b$, we can see that the dimension of the interior small square is $b - a$. Thus, we see that the area of the small square is $(b - a)^2 = b^2 - 2ab + a^2$. The area of each triangle is $(1/2)ab$, which means the four of them together have area $2ab$. Adding the triangles and small square together yields $a^2 + b^2$, and this equals the area of the larger square, which has area $c^2$. Hence, $a^2 + b^2 = c^2$.

Behold!

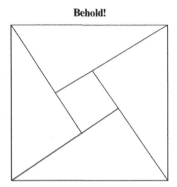

**FIGURE 8.1**   Bhaskara II's Proof of the Pythagoream Theorem: "Behold!"

A final note about Bhaskara II is a legend that involves the wedding of his daughter. Bhaskara II used astrological calculations that told him his daughter's soon-to-be husband would die if they were not married at the correct time. Bhaskara II set up a water clock so that the wedding could occur at the necessary time. Unfortunately, Bhaskara's daughter was curious, and while looking at the clock, a pearl from her nose ring fell into the water. This upset the workings of the clock and Bhaskara's daughter was married at the wrong time. Her husband subsequently died soon after the wedding. Since widows were not permitted to remarry, Bhaskara II taught his daughter mathematics.

Bhaskara II was the last great mathematician of his time. He died in 1185 CE. There was a significant gap in Indian mathematics achievement until the Kerala period in the 14th century CE. Madhava, founder of the Kerala School of Astronomy and Mathematics, was born near Kerala in southwestern India around 1350 CE and died in 1425 CE. Madhava did work in algebra, geometry, trigonometry, astronomy, and calculus. The biggest contributions from Madhava and his school were in mathematical analysis, including the infinite series and expansion for trigonometric functions and π. Like Bhaskara II, Madhava contributed greatly to differential calculus, and Madhava and his school even worked in integral calculus. One of the more interesting aspects of the Kerala School is that prior to this time period, most people studied in their homes and used their family's private library. The Kerala School served as a centralized location for astronomical and mathematical study. The Kerala Period ended in the 16th century

CE, about the time the Mughal Empire controlled vast amounts of modern-day India, Bangladesh, Pakistan, and Afghanistan. An emperor of the Mughal Empire, Shah Jahan, was the one responsible for building the Taj Mahal in Agra in the 17th century CE, which was a mausoleum dedicated to one of his wives.

An Indian mathematician that would perhaps surpass any other Indian mathematician was Srinivasa Ramanujan. He was a Pythagorean in the vegetarian sense and was born in 1887 CE in southern India. Ramanujan had very little training in formal mathematics but was able to derive many results that took centuries to develop in other parts of the world, such as a method for solving quartic equations. Ramanujan also independently made discoveries that have been found by European mathematicians such as the Bernoullis and Gauss. He also did work in analysis and infinite series, and died in 1920 CE at age 32 from poor health. We could speculate how much more Ramanujan could have done had he been given the support and access to the work of other mathematicians, as well as living a longer life.

There is an interesting method of multiplication that was most likely developed in India. Similar to other developments made in Indian mathematics, this procedure was brought to Europe from India by Islamic scholars. Indeed, it was unknown to Europe until the 13th century CE. The lattice method of multiplication, like the Egyptian method, is taught today in US classrooms in the *Everyday Mathematics* curriculum. If we would like to multiply 56 by 32, we can create a lattice as follows:

```
    5    6
+---+---+
|  /|  /|
| / | / | 3
|/  |/  |
+---+---+
|  /|  /|
| / | / | 2
|/  |/  |
+---+---+
```

Next, we multiply each single digit number by its corresponding single digit number to fill in the boxes. For example, $5 \times 3 = 15$, so the top left box gets 15 with the diagonal separating the 1 and the 5.

```
    5    6
  +---+---+
  |1 /|1 /|
  | / | / | 3
  |/ 5|/8 |
  +---+---+
  |1 /|1 /|
  | / | /| | 2
  |/ 0|/ 2|
  +---+---+
```

Next, we add through the diagonals as follows. For the first digit in the bottom far right, we simply bring down the 2. Along the next diagonal, we see that $0 + 1 + 8 = 9$. We continue this process, and finally we see that our answer is 1792.

```
      5    6
    +---+---+
    |1 /|1 /|
  1 | / | / | 3
    |/ 5|/8 |
    +---+---+
    |1 /|1 /|
  7 | / | / | 2
    |/ 0|/ 2|
    +---+---+
      9    2
```

The reason this method works is that the diagonals keep track of the place value. This could also work for numbers with more than two digits. However, sometimes carrying is necessary. See the example illustrated for 56 multiplied by 39 to yield 2184. Notice then when carrying is necessary, we carry to the next diagonal. For example, $5 + 5 + 8 = 18$, so the 10 from the 18 carries as a 1 to the next diagonal.

```
      5  1 6  1
    +---+---+
    |1 /|1 /|
  2 | / | / | 3
    |/ 5|/8 |
    +---+---+
    |4 /|5 /|
  1 | / | / | 9
    |/ 5|/ 4|
    +---+---+
      8    4
```

A final note on Indian mathematics is a discussion of the Tower of Brahma problem, which is sometimes known as the Tower of Hanoi problem after the Flag Tower in Hanoi, Vietnam. This problem is a mathematical puzzle created by French mathematician Edouard Lucas in 1883. The problem states there are three pegs on a piece of wood. On one peg, disks are stacked in decreasing size, which means the larger disks are at the bottom and the smaller ones are at the top. The goal is to move all of the disks from one peg to another, moving only one disk at a time in as few moves as possible. One is permitted to place smaller disks on top of larger disks, but one is not allowed to place larger disks on top of smaller disks. This would require $2^n - 1$ moves. For example, if we have three disks, we could do this as efficiently as possible in $2^3 - 1$, or 7, moves. The legend says there is an ancient temple in India where priests move giant golden disks from one large peg to another, and when the task is completed, the world will come to an end. There are 64 golden disks in the temple that must be moved. How long will it take for the world to end if the priests move one disk every second? If the priests began this process 3000 years ago, are we in any danger of the world ending soon, according to the legend? It would at first appear that our end is near since $2^{64} - 1$ does not intuitively seem very large. However, the answer turns out to be 18,446,744,073,709,551,615 seconds, which is about 585 billion years. We are in no danger of the world ending anytime soon, or at least we are in no danger of the world ending because of the priests completing their task in the temple.

Indian mathematics made many great contributions and likely influenced much of the European mathematicians to follow. History has not given Indian mathematics the credit it deserves since many European mathematicians did not give credit to the Indian mathematicians who helped them in their discoveries. As is discussed in the next chapter, Indian discoveries were transmitted to Islamic mathematicians, who then brought this knowledge to Europe. Thus, many European innovations come directly from Indian influence by the way of Islamic mathematicians.

# CHAPTER 9

# ISLAMIC MATHEMATICS: PRESERVATION, INNOVATION, AND CONNECTION BETWEEN EAST AND WEST

Islam traces its beginning to its founder, Muhammad, who was born around 570 CE in Mecca, which is in modern-day Saudi Arabia. At the age of 40, Muhammad, regarded by Muslims as the last prophet, retreated to meditate, and Muslims believe he began receiving revelations from God. These were recorded in what is now the Qur'an. Islam means "surrender," based on Muhammad's claim that followers must surrender to God, who is called Allah. Islam had already spread across the Arabian Peninsula by the time Muhammad died in 632. After Muhammad's death, the Islamic caliphate, or Islamic rulers, continued to spread Islam over the next 100 years. Islam reached East throughout Persia, toward India, West into Egypt, across North Africa, and finally into Spain. It was at the Battle of Tours in central France in 732 that Charles Martel, grandfather of Charlemagne, the first emperor of what developed into the Holy Roman Empire, stopped the Islamic advance into Europe.

Islam entered a Golden Age between 750 and 1258 with great advancements in philosophy, literature, medicine, science, and

*The Development of Mathematics throughout the Centuries: A Brief History in a Cultural Context*, First Edition. Brian R. Evans.
© 2014 John Wiley & Sons, Inc. Published 2014 by John Wiley & Sons, Inc.

mathematics. There was strict orthodoxy during the early stages of Islam. However, after its first century, there was an acceptance of academic rigor outside religious studies alone. It was during this time that great accomplishments were made in education, including the world's first university, which was Al-Azhar University in Cairo in the late 10th century. For most of this time, Baghdad was the center of the Islamic world. Beginning in the 11th century and continuing until the 13th century, European Christians fought the Crusades against the Islamic world in order to capture Jerusalem and other areas of the Holy Land. The Golden Age of Islam ended at the Siege of Baghdad in 1258 when Mongol invaders destroyed the city.

Islamic mathematics began after the start of the Golden Age around the beginning of the 9th century. A succession of ruling caliphs encouraged learning and advancement, which led to a rich culture of mathematics during the Islamic Golden Age. While Christian Europe disregarded Greek mathematical works, the Islamic world remained busy translating the works of the ancient Greeks into Arabic and studying their own mathematics. Along with Greek mathematics, the great philosophical works, such as those of Plato and Aristotle, were translated into Arabic. At that time, Arabic was the academic language in the Islamic world, just as Latin was the academic language in Europe. In the early 9th century, Al-Hajjaj ibn Yusuf Mater translated Euclid's *Elements* and Ptolemy's *Geographia* into Arabic. Around the same time, the works of Plato and Aristotle were translated into Arabic by Hunayn ibn Ishaq. It is because of Islamic mathematicians that we still have access to many of the great ancient Greek works.

Islamic mathematics is notable for its contributions to arithmetic, algebra, geometry, trigonometry, and calculus. Islamic mathematicians, influenced primarily by Babylonian, Greek, and Indian mathematics, had great influence on European mathematics in the High and Late Middle Ages and into the Renaissance and later centuries. As stated earlier, Egyptian and Babylonian mathematics influenced Greek mathematics, and Greek and Indian mathematics influenced Islamic mathematics, which spread its influence to Europe. European mathematics entered a dark period after the fall of the Roman Empire until the High and Late Middle Ages and the Renaissance, but while

Europe was in the dark, Islamic mathematics continued to progress. Modern mathematics has many similarities to the mathematics of the Islamic scholars.

The most important Islamic figure in mathematics from this period was Persian mathematician Muhammad ibn-Musa al-Khwarizmi. He was born around 780, and along with Diophantus he is also called the "Father of Algebra." He worked at Bayt al-Hikma, or "House of Wisdom," which was the academic center of the Islamic world established by caliph al-Ma'mun in Baghdad from the 9th to 13th centuries until its destruction in 1258 during the Mongol invasion. The House of Wisdom is where many of the translations of the ancient Greek works took place.

In the early 9th century, al-Khwarizmi wrote *Hisab al-jabr walmuqabala*, or *The Book of Calculation by Completion and Balancing*. In the 12th century, it was translated into Latin and called *Liber Algebrae et Almucabola*. This work gives us the word "algebra" from "al-jabr," which means "completion" in reference to solving for an unknown variable. This work is the basis for our modern concept of algebra and focused on linear and quadratic equations using practical applications. It should be noted that there was still a lack of mathematical notation by this point, which is the reason much of the algebraic work is written in words and not mathematical symbols. *The Book of Calculation by Completion and Balancing* represents a turning point in mathematics away from the Greek geometric view of mathematics, and from one perspective, it can be considered for algebra what the *Elements* is for geometry.

*The Book of Calculation by Completion and Balancing* addressed equations of six types, which involved solving by the use of squares, roots, and numbers. Only positive solutions, not negative solutions, were considered. In the representation to be shown later, note that modern mathematical notation is used instead of only the words used by al-Khwarizmi.

1. Squares can be equal to roots: $ax^2 = bx$.
2. Squares can be equal to numbers: $ax^2 = c$.
3. Roots can be equal to numbers: $bx = c$.

**4.** Squares and roots can be equal to numbers: $ax^2 + bx = c$.

**5.** Squares and numbers can be equal to roots: $ax^2 + c = bx$.

**6.** Roots and numbers can be equal to squares: $bx + c = ax^2$.

Another important work by al-Khwarizmi written around the same time was his *Book on Addition and Subtraction According to the Hindu Method*. This work was translated into Latin from the first lines of the text as *Algoritmi de Numero Indorum*, which means "al-Khwarizmi on the Hindu Art of Reckoning." It is from al-Khwarizmi's name that we derive the word "algorithm," which is any efficient method of calculation. This work is very important because it was the primary influence on bringing the Hindu–Arabic numeration system to Europe, along with the work of Arab mathematician al-Kindi around the same time. It should be noted that Europeans first called this system Arabic numeration because it was given to them through the Islamic works. However, today the system is called the Hindu–Arabic system due to the origin of the system in India and the refinement and transmission of the system by the Islamic world. It should be noted that there were no decimal fractions in this system; it would take several more centuries for Islamic mathematicians to introduce decimal fractions.

If the reader has lived or traveled in modern Arabian countries, he or she might be surprised that while the world uses the Hindu–Arabic system, it seems that one of the few places in the world that has symbols different from what we can call the Western Hindu–Arabic numerals is in the Arabian world. The Arabic symbols for 0 to 9 are used in many Arabic countries, but sometimes presented alongside the symbols most recognizable as Western Hindu–Arabic numerals. The Arabic numerals are sometimes called the Eastern Hindu–Arabic numerals and derive directly from written Arabic, while the Western Hindu–Arabic numerals derive from Indian numerals. As far as the positional number system is concerned, there is no difference between the two systems. The only real difference is the symbols that are used to represent the numbers 0 to 9 (Table 9.1).

**TABLE 9.1   Western and Eastern Hindu–Arabic Numerals**

| Western symbol | Arabic name | Eastern symbol |
|---|---|---|
| 0 | Sifr | ٠ |
| 1 | Waahid | ١ |
| 2 | Eeth-Nayn | ٢ |
| 3 | Thalaatha | ٣ |
| 4 | Arba'a | ٤ |
| 5 | Khamsa | ٥ |
| 6 | Sitta | ٦ |
| 7 | Sab'a | ٧ |
| 8 | Thamaaneeya | ٨ |
| 9 | Tis'a | ٩ |

Also around the time of his other two famous works, al-Khwarizmi had a third work of major importance, which was a revised version of Ptolemy's *Geographia* called *The Image of the Earth*. This work corrected mistakes made by Ptolemy, and contained calculations with longitudes and latitudes. A major correction of Ptolemy's work was the representation of the Atlantic Ocean and Indian Ocean as open bodies of water rather than large "lakes" surrounded by land, as Ptolemy had depicted. Al-Khwarizmi died in 850.

In the same time period as al-Khwarizmi, there were three Persian brothers called Banu Musa, which means "sons of Moses," who also worked in the House of Wisdom. Their most famous work was *The Book of Measurement of Plane and Spherical Figures*, which continued the work of Archimedes from his *Measurement of a Circle*. There is a story that says there was a dispute between the Banu Musa brothers and al-Kindi, who was mentioned earlier for his early work in the Hindu–Arabic numeral system alongside al-Khwarizmi's work. The story says that due to the Banu Musa brothers' influence, the caliph had al-Kindi beaten, and subsequently al-Kindi's library was given to the Banu Musa brothers. Al-Kindi was also remembered for his philosophical work and is called "The Arab Philosopher." He was born in 801 and died in 873.

In the 9th and 10th centuries, an Arab mathematician named Abu Kamil ibn Aslam from Egypt made innovations in solving systems of equations in three variables with three unknowns. Born around 850, he was also the first to use irrational numbers as roots to algebraic equations. Abu Kamil, like other Islamic mathematicians of his time, felt that his algebraic proofs should be grounded in geometry in the ancient Greek tradition. He advanced some of the work of al-Khwarizmi. His work was highly influential in introducing Islamic mathematics to Europe, and this is particularly evident in his influence on Leonardo of Pisa's 1202 work, *Liber Abbaci*. It is generally accepted that *Liber Abbaci* introduced Europe to the Hindu–Arabic system. Even though it was not the first work in Europe to present the Hindu–Arabic system, it was the most significant. Abu Kamil died around 930.

During the time of al-Khwarizmi, the Islamic world was using the Indian numeration system. However, decimal fractions were not yet in existence. In the 10th century, fractions and decimal point notation were found in the work of Abu'l-Hasan al-Uqlidisi, an Arabic mathematician who worked in Baghdad and Damascus. His most famous work that provides evidence of this is the *Arithmetic of Al-Uqlidisi*. He was born around 920 and died around 980.

Jamshid al-Kashi was born in Kashan, in what is now central Iran, around 1380. Al-Kashi improved on decimal fractions, giving us our modern notation for expressing fractions. Some consider al-Kashi to be the mathematician who completed the Hindu–Arabic system. He not only developed the law of cosines but also created an incredibly accurate approximation for $\pi$, one that would not be improved for another 200 years. Al-Kashi died in 1429.

Persian mathematician Muhammad ibn al Husayn al-Karaji was born in 953 in Baghdad and died around 1030. Al-Karaji grounded algebra in arithmetic and not geometry. He wrote *Al-Fakhri*, or *Glorious on Algebra*, which allows us an early look at mathematical induction. It was further developed in the following centuries by Ibn al-Haytham and Samau'al al-Maghribi. Mathematical induction is a form of mathematical proof that means we first show something is true in the first possible case of a mathematical claim. Next, we assume what we want to show in a given case, called $k$, and then show that $k + 1$, the consecutive case, must be true. This means we essentially build on the first possible case by showing that if a previous case is true, the consecutive case will also be true. Hence, we can generalize our claim as being true and proceed forward as high as we want to go, even leading into infinity. Exhausting all cases might not be possible, such as for the infinite, which means that mathematical induction is an innovative way to prove a certain claim is true. Al-Karaji used induction for his work in the binomial theorem and Pascal's triangle. We shall now look at a simple example of mathematical induction.

Let us say we want to show that the following is true: the sum of the first $n$ positive integers is $n(n + 1)/2$. In other words, $1 + 2 + 3 + \cdots + n = n(n + 1)/2$. Our first step is to show that this is true for $n = 1$: $1 = 1(1 + 1)/2 = 2/2 = 1$. Our next step is to assume that this is true for $k$, and algebraically show this is true for $k + 1$. To assume that this is true for $k$ means: $1 + 2 + 3 + \cdots + k = k(k + 1)/2$. We need to show that the case for $k + 1$ is true, which means we replace our original $n$ with $k + 1$: $1 + 2 + 3 + \cdots + k + k + 1 = (k + 1)(k + 2)/2$. Since we have assumed that the claim is true for $k$, we know that the first $k$ integers on the left side of the equation can be replaced with $k(k + 1)/2$. This means our new left side of the equation is $k(k + 1)/2 + k + 1$. Finding the common denominator of 2 gives us $k(k + 1)/2 + 2(k + 1)/2$. Next, we add the two quantities together to get $[k(k + 1) + 2(k + 1)]/2$, which when factoring a common $k + 1$ in the numerator simplifies to $(k + 1)(k + 2)/2$. Since this equals the right side of the equation, we are finished with the proof, and we can end with *quad erat demonstratum*, since our claim has been shown.

Although Islamic mathematics is best known for its innovations in number and algebraic development, there was also substantial work done in geometry, trigonometry, calculus, and combinatorics. Most notably, Islamic mathematicians did early work in analytic geometry, which is the application of algebra to geometric problems. Ibn al-Haytham, also known as Alhazen, was a Persian mathematician born around 965. His work is considered to be early algebraic calculus. Alhazen, like many other mathematicians to come, attempted to prove Euclid's parallel postulate using the other postulates in the axiomatic system. He died in 1040.

In the late 11th century, Umar Khayyam, a Persian mathematician born in 1048, worked with cubic equations. He was the first to reject the parallel postulate and consider non-Euclidean geometries such as elliptical and hyperbolic geometry. He also worked with Pascal's triangle and calculated the length of a year with good accuracy. Khayyam was also known for his poetry work, the *Rubaiyat*, which means "four" in Arabic because it contains a collection of poems with four lines or quatrains. Khayyam is also remembered for his questioning of religion, a rare occurrence at that time in the Islamic world. He died in 1131.

The use of all six trigonometric functions, based on the work of Indian mathematicians, was found in the work of Abu al-Wafa Buzjani, a 10th-century Persian mathematician and astronomer. For Islamic scholars, trigonometric development was closely tied to astronomy. Islamic mathematicians worked with calculus more than 500 years before Europeans discovered it. A 12th-century Persian mathematician named Sharaf al-Din al-Tusi was the first to find the derivative of a cubic polynomial.

Indian mathematicians were interested in combinatorics, and Islamic mathematics continued these investigations. Islamic mathematicians were able to use induction to prove the formulas for permutations and combinations. Ahmad ibn Munim al-Abdari, from Morocco in the 13th century, proved the formula for permutations, and Ibn al-Banna al-Marrakushi, also from Morocco in the 13th and 14th centuries, proved the formula for combinations.

Islam's tolerant stance toward learning outside religious studies began to shift by the 11th century. Mathematics continued to flourish during this time, but as the religion became more orthodox, mathematics and science discoveries declined. Religious leaders began to discourage study that was not religious in nature, and institutions of learning focused more heavily on religious studies over the next several centuries. In addition to the more orthodox stance developing during this time, the House of Wisdom was destroyed in 1258 during the Siege of Baghdad by Mongol invaders. The Mongol invaders, formerly led by Genghis Khan before his death, were particularly brutal and slaughtered the people of Baghdad. They destroyed the city completely, including the canals and irrigation system for agriculture that allowed the region to prosper for millennia. By this point, Europe was pulling itself out of its dark times and began expanding on the great mathematical works of the Islamic world. The contributions from Islamic mathematicians were indeed great. They had improved vastly on Babylonian algebra and combined the practical nature of the subject with the rigor of proof found in Greek mathematics. Not only did they preserve the works of the ancients but they also added their own rigor and innovations that helped propel mathematics forward in Europe in the coming years.

Isolated from Africa, Asia, and Europe is the Americas. Even though their mathematics had no influence on Africa, Asia, or Europe, and would eventually be left to history nearly forgotten, it is worth examining for the sake of its beauty and sophistication. In the next chapter, we shall explore the mathematics of Pre-Columbian America.

# CHAPTER 10

# PRE-COLUMBIAN AMERICAN MATHEMATICS: THE OLMEC, MAYA, AND INCA CIVILIZATIONS

The Olmec civilization flourished from about 1500 to 400 BCE in Mesoamerica. The people settled around San Lorenzo in modern-day Mexico on the Gulf of Mexico. Today, they are remembered primarily for their colossal head sculptures. Some believe that the faces on some of the colossal heads provide evidence of contact with or origin in Africa. However, the vast majority of researchers reject this claim. It is believed that the Olmec civilization was the first to develop writing in the Americas. The Olmec people are believed to have played an ancient ball game that influenced the ball games of later civilizations in Mesoamerica, including the Maya and Aztec civilizations.

It is possible that the Olmec had developed a concept of zero as a placeholder before anywhere else in the world. There is evidence that the Long Count calendar used by the Maya was actually developed by the Olmec, and zero as a placeholder is a necessary idea in the Long Count calendar. However, there have been disputes on this since the oldest record of the calendar was found after the demise of the Olmec civilization. One of the earliest artifacts of the Long Count calendar is the Stela C, a rock slab found in 1939 in Tres Zapotes. Stela C indicates a date of 32 BCE with one of the earliest uses of zero as a placeholder in history. Around 400 BCE, the Olmec population declined considerably, which was likely due to problems with environmental change and subsequent agricultural productivity.

The Maya civilization developed about 4000 years ago primarily on the Yucatan Peninsula in modern-day Mexico, and also in modern-day Belize, Guatemala, Honduras, and El Salvador. However, significant development occurred during the Classical Period between 250 and 900 CE. A lasting accomplishment of the Maya is the stepped pyramids, such as the one at Chichén Itzá, built in the early 7th century CE. The Maya developed the most advanced writing system in Pre-Columbian America, and they made great advancements in astronomy and mathematics. Today the *Dresden, Madrid,* and *Paris Codices* are surviving documents from the Maya, which were written on a type of tree bark. *The Dresden Codex* originated around the 11th century CE and is the oldest book we have from the Americas. It was damaged in World War II, but has been restored to an extent.

The Maya had a base 20, or vigesimal, numeral system (Table 10.1). It is possible that the Maya used base 20 due to human hands and feet containing 20 fingers and toes, and the Maya word for "man" is used for 20 as well. The number 5 was important for grouping, and this may have been due to one hand having five fingers. However, the Mayan system was not a true base 20 positional system because it contained only three symbols, although it did rely on position to an extent. The symbol for the number 1 was a dot and the symbol for the number 5 was a bar. The Maya had a special symbol for zero that was based on a shell and was likely introduced by the early Classical Period.

Notice the logical progression from 1 to 19. The names of numbers after 19 were also reasonable. The name for 20 was "one-twenty" or in Mayan "hun-kal," or also known as "kal." The number 40 was named "ca-kal," which is to say "two-twenties."

Symbolically, the number 20 requires explanation. Similar to the Chinese numeral system, the Maya system is represented vertically by place for numbers 20 and onward. The dot represents one 20 and

**TABLE 10.1  Mayan Numerals**

| Number | Name | Symbol | Number | Name | Symbol |
|---|---|---|---|---|---|
| 0 | Xix-Im | ⬭ | 10 | Lahun | ══ |
| 1 | Hun | • | 11 | Tuluc | • ══ |
| 2 | Ca | • • | 12 | La-Ca | • • ══ |
| 3 | Ox | • • • | 13 | Ox-Lahun | • • • ══ |
| 4 | Can | • • • • | 14 | Can-Lahun | • • • • ══ |
| 5 | Ho | ─── | 15 | Ho-Lahun | ═══ |
| 6 | Uac | • ─── | 16 | Uac-Lahun | • ═══ |
| 7 | Uuc | • • ─── | 17 | Uuc-Lahun | • • ═══ |
| 8 | Uaxac | • • • ─── | 18 | Uaxac-Lahun | • • • ═══ |
| 9 | Bolon | • • • • ─── | 19 | Bolon-Lahun | • • • • ═══ |

the shell represents zero 1s. There is a space left between the dot and the shell.

•

⬭

We would write 23 as follows:

•

We would write the Mayan 67 as follows, which is three 20s and seven 1s:

•••

Adding and subtracting using the Mayan system was fairly straightforward. If we wanted to add 4 and 7, we would represent this as follows:

However, we would not use six dots, so this means we need to trade five dots for one bar and we would have the following, which is 11:

It is also possible that we may need to regroup for subtraction. If we want to subtract 4 from 11, we would have the following, which is 7:

This is equivalent to the following:

It should be mentioned that as logical as the Mayan system appears to be, there is an interesting aspect that does not seem to make much sense. In our base 10 system, we progress in place value as follows: 1, 10, 100, 1000, . . . . We would expect the Maya base 20 system to progress in place value as follows: 1, 20, 400, 160,000, . . . In other words, our system is $10^0$, $10^1$, $10^2$, $10^3$, . . . , and the Maya system should be $20^0$, $20^1$, $20^2$, $20^3$, . . . . Instead, the Maya system progresses as follows: 1, 20, 360, 7200, . . . . We can also represent this as $20^0$, $20^1$, $18 \times 20$, $18 \times 20^2$, . . . . In other words, in the third power of 20, we only have $18 \times 20$ and not the expected $20 \times 20$. However, the next power reverts back to multiples of 20. This strange aspect of the Mayan system very likely originated in calendar creation, an important purpose for the Mayan system. There are approximately 360 days in a year, so it was convenient to have 360 appear in the number system.

Mayan religion was heavily dependent on the calendar to determine ritual and ceremony times. The Mayan people believed that the night sky, and hence astronomy, was critical to understanding the supernatural. The Maya considered the Sun and the Moon to be gods who assisted them against darkness. It is possible that time itself was worshiped as a god, which should not be a surprise considering the importance of the calendar. Like other civilizations around the world, religion may have been a major motivation to develop astronomy and mathematics to inform calendar development.

The Mesoamerica Long Count calendar, which may have developed during the Olmec civilization, indicates that creation occurred in 3114 BCE. There were two calendars used by the Maya. One had 365 days in a year, called the Haab, and the other had 260 days in a year, called the Tzolkin. The Haab was used as a regular calendar such as for agricultural purposes, while the Tzolkin was a ritual calendar.

The Haab had 18 months, with 20 days in each month. The final five days appeared at the end of the year and were called the Wayeb. The Wayeb was a fearful time in which the supernatural world could pass into the natural one. The Tzolkin had 13 months, with 20 days in each month.

Approaching December 21, 2012, some people in the West believed that the Long Count calendar indicated the end of the world due to the coming end of a 5125-year cycle. However, this was a misinterpretation of the Long Count calendar. The Maya believed that there were cycles before the recent cycle, and there is nothing in Mayan literature to indicate a belief that the last cycle should have been the final cycle.

Mayan civilization declined and collapsed in the 8th and 9th centuries CE. The cause of both the decline and collapse is uncertain, but it may have been overpopulation or environmental changes affecting agriculture, as may have happened with the Olmec civilization. Over the next 600 years, the Mayan civilization lingered until the arrival of the Spanish, but it would not regain the glory of the Classical Period.

The Spanish first landed on the Yucatan Peninsula in 1511 by accident. Most of the crew were enslaved and killed by the Maya. However, because they brought smallpox to the Mayan civilization, the disease soon killed many. Hernán Cortes, the Spanish conquistador who also later defeated the Aztecs, arrived at the Yucatan Peninsula in 1519. By 1527, many Maya learned that Cortes and the Spanish had defeated the Aztecs in 1521. This prompted some Maya to pledge their allegiance to the Spanish king. However, Mayan resistance against the Spanish lasted for nearly 170 more years because, unlike the Aztecs, Mayan government was not as centralized so resistance could continue in several areas even after the fall of another political center.

The Inca civilization constituted the largest empire in the Pre-Columbian Americas and spanned from about 1200 to 1533 CE. The Inca Empire was centered around the Andean city of Cusco in modern-day Peru and also occupied parts of modern-day Ecuador, Bolivia, Chile, and even parts of Columbia and Argentina. Government during the Inca civilization was a theocracy. The spiritual leader, Sapa Inca, was also the ruler of the empire. The Sapa Inca was considered a descendant of the Sun god, called Inti. However, Inca civilization is probably most famous for its impressive architecture. Inca architecture did not use mortar to hold the stones together. Yet their building techniques were advanced enough to allow the stones to fit together perfectly. A lasting legacy of the Incas is the 15th-century Inca city near Cusco, Machu Picchu.

Another notable aspect of Inca life was its educational system. Instead of a writing system, the Inca used a device called the quipu, which means "knot." The quipu is a grouping of many strings with various knots to record numerical information. The top end of the string was held together at the center of the quipu. The number system used a positional base 10 system with simple knots representing powers of 10s such as $10^1$, $10^2$, $10^3$, . . . . Long knots represented units of 1, and the absence of a knot indicated zero. The number 1 was represented with a unique knot tied in a figure "8" loop. The position of the knots on the string was important in determining the number. The ones units were at the bottom of the string, and each power of 10 was evenly spaced moving toward the top of the string. Strings were of different colors to help make number identification easier. The accountants who used the quipu were called Quipucamayocs, which means "the authority of the quipu."

If we want to represent the number 64, we would have four long knots at the bottom of the string and six simple knots moving up the string, which is represented as follows:

—|—|—|—|————————I—I—I—I—I—I————————

If we want to represent the number 253, we could represent it as follows:

—|—|—|————————I—I—I—I—I————————I—I—

Finally, if we want to represent the number 203, we could leave a space where we have no 10s:

—|—|—|——————————————————————I—I—

After arriving in South America from Panama in 1526, Spanish conquistador Francisco Pizarro returned to Spain to gain permission to conquer the wealthy Incas, and he succeeded in doing so by 1533. Pizarro and his forces were vastly outnumbered by the Incas, but with Pizarro's advanced weapons and the new smallpox disease killing many Incas upon the Spanish arrival, Pizarro's victory was assured. Additionally, Pizarro captured the Inca leader, Atahualpa, and was able to control the Inca Empire directly through him. Eventually, Pizarro had Atahualpa executed, and this ended the Inca Empire.

While the mathematics of the Pre-Columbian Americas had no impact on the mathematics of Africa, Asia, and Europe, it was quite innovative and shows the genius of the indigenous population of the Americas. In the modern-day United States, it is important that teachers of mathematics emphasize the vast mathematical heritage of Latin America to encourage the many students who have their own heritage in that part of the world.

In the next part of the book, we shall explore the mathematics that developed in Europe beginning in the High and Late Middle Ages, the Renaissance, and into the 17th to 20th centuries. We shall see that many of the contributions from Indian and Islamic mathematics were an enormous influence on renewed European thought.

# MATHEMATICS IN EUROPE

MATHEMATICS IN
EUROPE

# CHAPTER 11

# EUROPEAN MATHEMATICS: THE MIDDLE AGES AND THE RENAISSANCE

After the fall of the Roman Empire in 476 CE, Europe entered the Middle Ages. However, the Byzantine Empire, or Eastern Roman Empire, centered in Constantinople continued until 1453 when Constantinople fell to the Muslim Ottoman Turks. Until recently, the early part of this period had been regarded as the Dark Ages, but the use of the term "Dark Ages" has generally fallen out of favor among historians due to its negative connotation. Historians have increasingly acknowledged accomplishments during this period, even though it is still comparatively limited compared with ancient Greece, Rome, or the Renaissance. The Dark Ages may not be as dark as commonly believed, considering the progress that was made, but in terms of mathematics, it was indeed a dark period in Europe with very little innovation. There continued to be an interest in using mathematics, but mathematics was studied for more practical purposes without a drive to make significant advancements. As in other cultures, people were motivated to study mathematics for the purposes of calendar making for religious observances. Examples include calculating the

*The Development of Mathematics throughout the Centuries: A Brief History in a Cultural Context*, First Edition. Brian R. Evans.
© 2014 John Wiley & Sons, Inc. Published 2014 by John Wiley & Sons, Inc.

date for Easter and using mathematics in architecture to build cathedrals and churches.

The Middle Ages is generally defined as the period between the fall of the Roman Empire in 476 CE and the beginning of the Early Modern Era around 1500. The Renaissance occurred between 1300 and 1600, which means that it was during the later part of the Renaissance that we designate the transition from the Middle Ages to the Early Modern Era. The Middle Ages can be subdivided into three general periods: the Early Middle Ages (476–1000), High Middle Ages (1000–1300), and Late Middle Ages (1300–1500). Christianity spread rapidly throughout Europe during the Early Middle Ages, and some scholars believe that this was one of the reasons that Europe's intellectual development stagnated. With Christianity's emphasis on the afterlife and the coming Armageddon, there was no urgency or need to study subjects such as mathematics, which meant that innovations in mathematics were not occurring in Europe during the Early to High Middle Ages. Many believed the world would end by 1000, but when it did not, many Christians realized that humans would be here longer than they had initially thought. By the 11th century, universities were being established, first the University of Bologna in Italy, and shortly after, Oxford University in England. Moreover, in the 13th century, Catholic philosopher Thomas Aquinas took a big step forward when he embraced reason outside faith as a means of gaining knowledge and accepted the methods of Aristotle. Until this time, many Christian thinkers had rejected non-Christian philosophers and considered only God's revelation as a means to attaining truth.

While much of Greek thought diminished in Europe during this period, some things remained intact. The Greek and Roman quadrivium continued and mathematics was studied under the great influence from the works of Boethius. However, it is important to note that no real innovations took place. Most of the education involving the use of mathematics was inherently superficial during the Early Middle Ages. As we have seen in previous chapters, innovative mathematics was occurring rapidly in China and India. During the end of the Early Middle Ages to the Late Middle Ages, Islamic mathematics was in its Golden Age.

It was during the High Middle Ages in the 12th century that translations of the major Greek works, as well as unique Islamic innovations, from Arabic to Latin occurred. Since the 8th century, Muslims occupied parts of the Iberian Peninsula, called Al-Andalus. This contact allowed European scholars to translate the Greek and Islamic works into Latin. It should be noted that Spanish Jews were instrumental in these translations because many of them were fluent in Arabic and Latin.

A prominent Spanish Jewish mathematician was Abraham bar Hiyya, who was born around 1070 in Barcelona and died around 1140. His best-known work is the *Treatise on Mensuration and Calculation*, which was translated into Latin in 1145. His work, influenced by the work of al-Khwarizmi, brought Islamic algebra to Europe and gave a complete solution to the quadratic equation. His work was significant given his methods of algebraic proof done in the Islamic tradition. His work was also based on geometric foundations.

The most important European figure during this time period was Italian mathematician Leonardo of Pisa, also known as Fibonacci, which means "son of Bonacci." He was born around 1170 and lived until around 1250. He studied under Islamic mathematicians in North Africa as a young man, and in 1202, Leonardo published his greatest work, *Liber Abbaci*. It is generally accepted that *Liber Abbaci*, or the *Book of Calculation*, introduced Europe to the Hindu–Arabic system, even though it was not the first work in Europe to address this topic. In 976, the *Codex Vigilanus* introduced the Hindu–Arabic system, but *Liber Abbaci* was much more influential. *Liber Abbaci* was influenced by the work of Abu Kamil in the 9th and 10th centuries, and it was Abu Kamil who had expanded on the work of al-Khwarizmi. Leonardo was also influenced by the work of Abraham bar Hiyya. *Liber Abbaci* was the best summary of Islamic mathematics in Europe at the beginning of the 13th century, but it covered Islamic mathematics only through the 10th century, even though Islamic mathematics had advanced during the 11th and 12th centuries. In addition to introducing the

Hindu–Arabic system as an easier system for calculation, *Liber Abbaci* also included the lattice method of multiplication and al-Khwarizmi's six types of equations. Other topics covered in *Liber Abbaci* were currency conversion problems, proofs from the *Elements*, systems of linear equations, and motion and mixture problems.

Leonardo presented an interesting problem for us to consider here. If the first person gives the second person 1 unit of currency, then the two will have the same amount. If the second person gives the first person 1 unit of currency, then the first person will have 10 times the amount that the first person has. How much money do they currently have? We can represent this algebraically by indicating that the first person has $x$ amount of money and person two has $y$ amount of money. This means that $x - 1 = y + 1$ and $10(y - 1) = x + 1$. We thus have two systems of equations that we solve by substitution by placing $y = x - 2$ into the second equation to yield $10[(x - 2) - 1] = x + 1$. Solving for $x$ we yield $x = 31/9$ or $3\frac{4}{9}$ and $y = 1\frac{4}{9}$, which means the first person had $3\frac{4}{9}$ units of currency and the second person had $1\frac{4}{9}$ units of currency. However, it should be noted that Leonardo solved this problem in a somewhat more complicated manner by introducing a third variable to represent the total money the two people had.

*Liber Abbaci* introduced the Fibonacci sequence to Europe, which appears as the following problem that was influenced by Indian sources. Imagine there are two rabbits that were just born, one male and one female. Rabbits can mate after turning 1 month old and the pregnancy will last for another month. This means a rabbit can produce offspring as early as 2 months after she is born. Let us assume our rabbits mate after the first month and produce two rabbits that will also be male and female after another month. Our original two rabbits will keep producing two babies, one male and one female, every month onward, infinitely. How many pairs of rabbits will there be after 1 year from the start of this? What is the pattern?

To answer this question we need to realize that each pair of rabbits depends on a previous pair of rabbits. We can say that the original pair of rabbits was born on July 31, which means that they mate on August 31. However, their offspring will not be born until about Sep-

**TABLE 11.1  Fibonacci Sequence**

| End of month | 0 | 1 | 2 | 3 | 4 | 5 | 6 | 7 | 8 | 9 | 10 | 11 | 12 |
|---|---|---|---|---|---|---|---|---|---|---|---|---|---|
| Number | | 1 | 1 | 2 | 3 | 5 | 8 | 13 | 21 | 34 | 55 | 89 | 144 | 233 |

tember 30. This means at the end of month zero, July, we have one pair of rabbits. At the end of month 1, August, we still have one pair and the female is now pregnant. At the end of month 2, September, we now have two pairs of rabbits because the original pair has two offspring. We assume when the parents give birth on September 30, they become pregnant again later that day. This means that on October 31 the parents gave birth again, and their offspring have become pregnant. Now there are three pairs of rabbits. This pattern continues as shown in Table 11.1, so that we have 233 pairs of rabbits at the end of the year, or July 31, one year after the original rabbits were born.

The Fibonacci sequence appears in nature, such as in the florets of many flowers including the sunflower. It also appears on the outside of a pineapple, in the pine cone, and in the shell of an African snail. Even the mating structure of honeybees can be represented with Fibonacci numbers.

In the early to mid-13th century, German mathematician Jordanus de Nemore, a figure we know very little about, worked in arithmetic, algebra, and geometry. It is possible that he taught mathematics in Paris and produced many mathematical works. Although his work did not expand much on the work of Islamic mathematicians, his most important contribution was his use of letters for unknown numbers. Jordanus visited the Holy Land around the mid-13th century, but unfortunately he died at sea en route back to Europe.

A European innovation in mathematical induction was introduced by a French Jewish mathematician, Levi ben Gerson. He was born in 1288 and he is known for his 1321 work, *Maasei Hoshev*, or the *Art of the Calculator*, in which ben Gerson expanded on the mathematical induction of the Islamic mathematicians. Furthermore, ben Gerson addressed finding square roots and cube roots and did extensive work in combinatorics and astronomy. He died in 1344.

William of Ockham, a Franciscan friar who was also a logician, was born in southern England in the same year as ben Gerson. He developed the idea of Ockham's razor, which states that the explanation that makes the fewest new assumptions should be the explanation adopted with all else being equal. The idea of the razor is to "cut away" the unnecessary assumptions. Ironically, despite the fact that this development comes from a Franciscan friar, the argument was later used as an argument against God's existence since the God hypothesis is an extra assumption in explaining the universe. William developed the logic of what would later be known as De Morgan's laws by the 19th century, which states that the negation of statement (*P* and *Q*) is (not *P* or not *Q*) and the negation of the statement (*P* or *Q*) is (not *P* and not *Q*). William died in 1348.

After centuries of little progress, Europe entered an age of accelerated advancement that began in the 14th century. The Renaissance, which means "rebirth," started in Florence, Italy, and continued throughout the 15th and 16th centuries. Although there were setbacks in European advancement due to the Hundred Years' War between France and England from the mid-14th to 15th centuries and the Black Death, which occurred between 1348 and 1350 and killed about a third to half of the population of Europe, the Renaissance period was a time in which people embraced the classical works and introduced advancements in art, literature, science, and mathematics. Art was an area that advanced rapidly during this period due to the application of mathematics used to expand artistic perception. The Renaissance spread throughout Italy in the 15th century and spread to Germany and then France shortly afterward. By the 16th century, the Renaissance arrived in England, around the time of William Shakespeare, and it eventually covered much of Europe.

There were several reasons for mathematical and scientific development during the Renaissance. First, after the year 1000, the world did not end as many expected, which led to interest in knowledge and nonreligious study because people began to believe that the world may last longer than expected. Furthermore, in the 13th century, Thomas Aquinas accepted reason as a means of investigation as opposed to relying solely on God's revelation. Second, foreign trade for the Italian city-states, which also generated great prosperity, became more reliant on mathematics in their transactions. A move from Roman numerals to the Hindu–Arabic system made calculations easier and led to more mathematical advancement in Europe. Italy was comprised of city-states on the Mediterranean Sea, just as was ancient Greece. Perhaps this configuration led to substantial advancements and affluence in both of their cases. Third, Johannes Gutenberg, born in Germany in 1398, produced the printing press in 1440. Over the next several centuries, this led to an incredible dissemination of knowledge not only in mathematics but also in all areas. It became more profitable to produce publications for authors, and more people now had access to information. The mathematics and scientific communities could more easily communicate with each other through the circulation of publications. Francis Bacon, philosopher and scientist born in 1561 and who is called the "Father of the Scientific Method" and the "Father of Empiricism," said that printing, as well as gunpowder and the compass, changed the world. Recall that gunpowder, the compass, and printing, in a more limited form, were also found in China well before Europe. The Protestant Reformation, which began with Martin Luther in Germany in 1517, gained much of its momentum from the availability of Protestant literature for the public. Information moved more freely and ideas could be built on the ideas of others. In a similar manner, the Internet and technology in the early 21st century are also leading a revolution in knowledge dissemination, which may advance human knowledge even more profoundly than the printing press. Fourth, the Ottoman Turks overran the Byzantine Empire in 1453, and many of the Byzantine scholars moved to Italy, along with their collections and knowledge of Greek mathematics and philosophy. This led to resurgence in interest in the classical Greek works.

Two figures who embodied the Renaissance were Leonardo da Vinci and Michelangelo Buonarroti. Leonardo was born near Florence in 1452. He is often considered to be the most talented person who ever lived, and the embodiment of the "Renaissance Man," or polymath, due to his achievements in art, architecture, music, science, mathematics, and many other areas of study. He may be best known for his paintings, including the *Mona Lisa* and *The Last Supper*, in addition to his drawing, the *Vitruvian Man*, which depicts Leonardo's conception of the ideal geometric human proportions using the golden ratio, which he called the divine proportion. Leonardo was a Pythagorean in that he was a vegetarian. Leonardo also did some work in mathematics. Near the end of the 15th century, Leonardo studied under Italian mathematician Luca Pacioli, who created mathematics textbooks and translated the *Elements* into Latin. Leonardo drew the illustrations for Pacioli's work in Platonic solids. Pacioli was one of the last of the abacists. Abacists had been around since the beginning of the 14th century in Italy and were mathematicians who wrote the mathematics textbooks used for instruction. The textbooks were used by the children of merchants in their education so that the children could learn sufficient mathematics for their trade.

A legend says that when Leonardo died in 1519 it was in the arms of French king Francis I. It was Francis I who had bought the *Mona Lisa*, which is now owned by the French government and housed in the Louvre Museum in Paris. Interestingly, considering Leonardo's magnificent accomplishments, his last words in the king's arms were alleged to have been that he had offended God and humankind because his work did not reach the quality he thought it should have reached. However, this may be more of a legend than fact.

Michelangelo was born in 1475 in Arezzo, which is southeast of Florence. He is considered to be the main rival for Leonardo as the embodiment of the "Renaissance Man." Michelangelo is well known for his art that depicted biblical imagery, such as his sculptures the *Pieta*, completed around 1500 and representing Mary and Jesus, and the *David*, representing King David from the Bible. Between 1508 and 1512, Michelangelo was commissioned to paint the ceiling of the Sistine Chapel in the Vatican, which would lead to one of his most famous works, the *Creation of Adam*. He was one of the architects for St. Peter's Basilica, and designed its famous dome. However, Michelangelo died in 1564 before the basilica was completed.

The Age of Exploration began in Europe at the early part of the 15th century with the Portuguese exploration of Africa's west coast in 1418. Christopher Columbus sailed to the Americas in 1492, believing he was finding a shorter route to Asia. We have already discussed the exploits of Cortés and Pizarro in Chapter 10. Vasco da Gama reached India in 1498 by sailing around Africa. In 1522, Ferdinand Magellan's expedition became the first to sail around the world. After his death in the Philippines, the expedition was led by his second-in-command, Juan Sebastian Elcano. It was during this time of European exploration—unfortunately, in many cases exploitation of native peoples—that the explorers required mathematics and astronomy for navigation, especially trigonometry, which appeared in Europe's first serious book in trigonometry, *On Triangles of Every Kind*, by Johannes Muller, also known as Regiomontanus. *On Triangles of Every Kind* was written in 1463, but it was not published until 1533. Regiomontanus also computed lengthy trigonometric tables.

The 16th century saw much advancement in mathematics. Girolamo Cardano was born in 1501 in Milan, Italy. Cardano's father was a lawyer who also worked in mathematics and taught him mathematics, and Cardano's father had advised Leonardo da Vinci in geometry and also lectured on this subject at the university. Girolamo Cardano studied medicine, and while being a brilliant student, he was not well liked at the university. In order to earn money, Cardano became involved in gambling, an area in which he was quite good. This led him to write the first mathematical book on probability called *Liber de Ludo Aleae*, or the *Book of Games and Chance*, which, although written in 1526, was not published until 1663. His greatest mathematical work was *Ars Magna*, or *The Great Art* (1545), which is the first algebraic work in Europe to advance beyond Islamic algebra. It contains long-sought-after solutions to cubic and quartic equations. The result on cubic equations came from the work of Niccolo Fontana, also known as Tartaglia, while the result on the quartic equations was discovered by his student, Lodovico Ferrari. Cardano was the first to develop the idea of an imaginary number, which today is symbolized by $i$ and is equal to the square root of $-1$. Around 1550, Cardano's son poisoned his own wife, whom Cardano did not like, and was eventually jailed and executed, which pained Cardano greatly. It was said that Cardano predicted the date of his own death, but this date could be due to Cardano possibly committing suicide in 1576.

Let us look at imaginary numbers. We symbolize the imaginary number $i$, and this is equal to the square root of $-1$. This means if we square $i$, we get $-1$. If we cube $i$, we get the negative square root of $-1$, which would be $-i$. If we take $i$ to the fourth power, we get 1 because this is $i^2$ multiplied by $i^2$, which is $-1 \times -1$. The remarkable property is we find $i$ to any exponent by considering these first four cases. If we have $i^5$, we can write this as $i^4 \times i$ and see that we have $-1 \times i$ to yield $-i$. This procedure works for any exponent. For example, if we have $i^{33}$, we can write this as $i^{32} \times i$, and since $i^{32} = i^{4 \times 8}$, we know that $i^{32} = 1$. Hence, our result is simply $i$. In general, if we want to simplify $i$ to any exponent, we look for multiples of 4 and simply observe what we have left.

Tartaglia, a contemporary of Cardano, was born in northern Italy around 1500. The name "Tartaglia" means the "stammerer"; when a French soldier stabbed him in the jaw during an attack on Tartaglia's town in 1512, a speech impediment resulted. Tartaglia, who later wore a beard to cover his scars, grew up poor and taught himself mathematics. He discovered how to solve the cubic equation and was convinced by Cardano to reveal his method. Tartaglia reluctantly revealed his solution with a promise from Cardano that he would not publish the result. However, while Cardano at first kept his promise, he later believed that this work had been done earlier by Italian mathematician Scipione del Ferro, and thus Cardano published the *Ars Magna* in 1545, giving Tartaglia some credit for the discovery. Tartaglia was outraged by the publication of his work and may have bribed Cardono's son to accuse him of heresy in 1570. Today, the solution to the cubic equation is called Cardano's formula, but sometimes referred to as the Cardano–Tartaglia formula. Tartaglia died in 1557.

Rafael Bombelli was born in 1526 in Bologna, Italy, and died in around 1572. Although it can be argued that Cardano began the process, Bombelli is best known for establishing imaginary and complex numbers, including using complex numbers as solutions to quadratic equations. Bombelli was also the first to state how to perform calculations with negative numbers. He is notable for his early use of mathematical symbols and notation.

French mathematician Francois Viete, born in 1540 and died in 1603, wrote the *Analytic Art* in which he improved on the use of mathematical symbols and notation. Viete used letters to indicate unknown quantities with vowels representing unknown variables and consonants representing constants. English mathematicians William Oughtred, born in 1573 and died in 1660, and Thomas Harriot, born in 1560 and died in 1621, expanded on Viete's work by creating a symbolic system of algebra.

Addressed in Chapter 9, Simon Stevin, born around 1548 in Belgium and died in 1620, fully developed the quadratic formula in 1594. Stevin did not invent the notion of decimals in the Hindu–Arabic system because decimals were used by the Chinese and Islamic mathematicians, but he was the first to introduce them to Europe in his work *Art of Tenths* in 1585. His work influenced Thomas Jefferson to adopt a decimal currency system for the United States in the 18th century. The notation used by Stevin would be improved on by Scottish mathematician John Napier, who was born in 1550 in Edinburgh. Napier discovered the logarithm, which is the inverse of the exponential function and highly useful in efficient computing in a time before calculators and computers. Logarithms are still useful in measuring earthquakes, and the logarithmic spiral appears in nature, such as in the nautilus shell. Napier published a book that contained logarithmic tables in 1614 called *Description of the Wonderful Canon of Logarithms*. Napier also aided in simplifying calculations by using "Napier's bones," which is essentially an abacus that is based on lattice multiplication addressed in Chapter 8. Napier's bones are a set of rods, originally made of ivory and appearing like bones, that were used on a board that allows a person to perform multiplication and division calculations fairly quickly. Napier was said to be interested in the Book of Revelation in the Bible, as well as the occult, and he believed the world would likely end before 1700. He died in 1617.

The Scientific Revolution began in 1543 with the publication of Polish astronomer Nicolaus Copernicus' *De Revolutionanibus Orbium Coelestium*, or *On the Revolutions of the Heavenly Spheres*, which claimed that the heliocentric model correctly explained the relationship between the Sun, the Earth, and other planets in the solar system. That is, the heavenly bodies, including the Earth, revolve around the Sun, and not around the Earth. This contradicted Ptolemy's geocentric model, which said everything revolved around the Earth and had been accepted for nearly 1400 years. Ptolemy's ideas were based on previous work of the Babylonians, which predated him by perhaps 800 years. Copernicus discovered there were proposals by the early Greeks, as far back as Aristarchus in the 3rd century BCE, which showed a heliocentric model.

Copernicus, born in 1473, was a polymath like Leonardo da Vinci. He excelled not only in astronomy but also in medicine, art, languages, politics, economics, and mathematics. Copernicus died in the year that his *On the Revolutions of the Heavenly Spheres* was published, so he avoided any religious controversy that would not be escaped by Galileo in the 17th century. Interestingly, the Catholic Church did not object to the book when it was published, but rather there was Protestant condemnation of the work, including the German founder of the Protestant Reformation, Martin Luther. It is possible that the Catholic Church did not condemn Copernicus' work because he did not state that his model was correct, but rather that his model made more sense mathematically. There is a legend that says Copernicus awoke from a coma long enough to have his newly published work placed in his hands so that he could look upon his published book before he died.

After Copernicus' death, Danish astronomer Tycho Brahe, born in 1571 and died in 1601, compiled astronomical data that were analyzed by his assistant, German astronomer Johannes Kepler. Brahe wore a prosthetic nose made of copper (sometimes it was silver or gold), after losing part of his nose in a duel. Born in 1571, Kepler formulated three laws of planetary motion, later called Kepler's laws. The first law states that the planets orbit the Sun in an elliptical manner, with the

Sun on one of the foci of the ellipse. The second law states that a line from the Sun to a planet creates equal areas in the ellipse over equal times. The third law states that for the planets, the ratio of the square of the time period of one revolution of that planet around the Sun to the cube of the average distance from the Sun is the same for all planets. Kepler was influential on Isaac Newton's law of gravitation. Kepler died in 1630.

Born in Pisa, Italy, in 1564, Galileo Galilei is considered, like Democritus in ancient Greece, to be the "Father of Modern Science," and he is also considered to be the "Father of Modern Physics." At 8 years old, his family returned to their former home of Florence, but Galileo joined them 2 years later when he was 10 years old. Galileo was educated at a monastery outside Florence. At first, he studied medicine, but then he switched to mathematics. Eventually, he would become the chair of mathematics at the University of Pisa, and after his father died in 1591, he would leave the University of Pisa for the University of Padua, which paid much more and helped him support his two sisters.

Galileo accepted the heliocentric model and expanded on the works of Copernicus and Kepler. In 1616, the Catholic Church condemned the work of Copernicus, and Galileo was instructed not to promote the heliocentric model. In 1632, he published his views in his work *Dialogue Concerning the Two Chief World Systems*, in which he contrasted the heliocentric and geocentric models. It was his heliocentric view in this work that put him before the Roman Inquisition. Galileo was convicted and forced to retract his statements in 1633. Although the Catholic Church had allowed Galileo to publish his work, he was required to present both the heliocentric and geocentric arguments, as well as the Pope's opinion. However, the Catholic Church felt that he violated this agreement and mocked the Pope in the process. The major opposition from the Catholic Church was that the heliocentric model contradicted the Bible. Psalm 104:5 says, "The Lord set the Earth on its foundation, it can never be moved." Also, Ecclesiastes 1:5 says, "And the sun rises and sets and returns to its place." According to the Catholic Church, the idea that the Sun was stationary and that the Earth moved directly contradicted these verses from the Bible. There is a legend that says after recanting his belief that the Earth revolved around the Sun, Galileo dramatically whispered under his breadth: "E pur si muove." This means: and yet it moves.

After his conviction by the Inquisition in 1633, Galileo was confined to house arrest for the remainder of his life and forbidden to publish any other works. Unfortunately, Galileo was not pardoned by the Catholic Church until over 350 years later in 1992 by Pope John Paul II. However, Galileo continued writing and secretly published his work. His major focus was the application of mathematics to physics with strict scientific method applied to his inquiries. His idea was that physics should be solidly grounded in mathematics, as he presented in his 1623 work, *The Assayer*. If a hypothesis did not hold up to mathematical scrutiny, it must be rejected and replaced with a better idea. Galileo brought science away from the teleological notions that everything in the world happens toward a religious end. The laws of nature are mathematical, and phenomena do not happen because they are caused directly by God for a purpose. This foundation in the scientific method and inductive reasoning would be one of the greatest contributions to science and mathematics by Galileo. In 1638, Galileo published his *Discourses and Mathematical Demonstrations Relating to Two New Sciences*, which examined the laws of motion and the strength of materials.

Galileo was very interested in the physics of projectiles and falling objects. Galileo proposed that falling objects, regardless of weight, had the same rate of acceleration, which is 9.81 m/s/s, which is in contrast to Aristotle's idea that objects fall at speeds related to mass. This means that a bag of feathers and a bag of rocks, when dropped from the same height at the same time, would accelerate at the same rate and thus hit the ground at the same time, disregarding wind resistance. It would be best to place the objects in a bag because a feather tends to float to the ground due to its shape and therefore would fall more slowly than the rock. A legend says that Galileo dropped two cannon balls of different masses from the Tower of Pisa to demonstrate his idea, but this is probably not true.

Galileo improved on the telescope, which was a new invention from The Netherlands in 1608. Due to his interest in pendulums and motion, Galileo designed a pendulum clock near the end of his life. Galileo died in 1642, the same year that Isaac Newton was born. Using the Gregorian calendar, Newton was born about a year after Galileo died, but at the time England was still using the Julian calendar. Newton expanded on Galileo's methods of inquiry and scientific discovery.

The Scientific Revolution rejected or refined many Greek scientific ideas, and generally advanced scientific progress outside religious influence, even though many of the thinkers during this period were religious people. The idea was that scientific inquiry could be conducted without reliance on religious canon. The Scientific Revolution continued into the 17th century with great advancements in science and mathematics. It could be argued that the printing press enabled the Scientific Revolution because the ideas generated were now more widely available to scientists and mathematicians. The new ideas could be built on, published, and read, and then more ideas could be added to the work. In the next chapter, we shall look at some of the great mathematicians and scientists of the 17th century and their accomplishments.

CHAPTER *12*

# EUROPEAN MATHEMATICS: THE 17TH CENTURY

The 17th century was perhaps the greatest century for mathematical advancements. It was during this century of accomplishment that the Scientific Revolution continued onward and people were able to see the world and universe in more ways than they ever had before, with the inventions of the microscope and telescope. Before the start of this century, William Shakespeare, perhaps the greatest writer in the English language, began writing some of the world's most famous plays such as *Romeo and Juliet* and *A Midsummer Night's Dream*. The major cities in the English North American colonies, such as New York, Boston, and Philadelphia, began to develop, and great colleges were founded in America, including Harvard in Massachusetts and William and Mary in Virginia. During this century, Oliver Cromwell temporarily overthrew the English monarch. On the European continent, the Thirty Years' War was waged between Protestants and Catholics in the Holy Roman Empire.

During this century was seen the great development of analytic geometry, probability, and calculus. As mentioned in the last chapter,

*The Development of Mathematics throughout the Centuries: A Brief History in a Cultural Context*, First Edition. Brian R. Evans.
© 2014 John Wiley & Sons, Inc. Published 2014 by John Wiley & Sons, Inc.

much of this advancement may have been due to the widespread dissemination of knowledge through the mass production of books and journals, which was enabled by the printing press. We begin our exploration of the 17th century with a very important mathematician and philosopher, René Descartes.

René Descartes, who was born in 1596 in the western part of central France in a town now named after him, is called the "Father of Modern Philosophy" for his great works in 17th-century philosophy. Along with Pierre de Fermat, Descartes developed analytic geometry, an area in mathematics that combines algebraic ideas on a geometric plane with equations representing curves; because of this, Descartes is called the "Father of Analytic Geometry." Recall that Alhazen, Islamic mathematician in the 10th and 11th centuries, had developed analytic geometry ideas centuries prior. The work of Descartes and Fermat established a foundation for the calculus of Isaac Newton and Gottfried Leibniz to come later in the century. In particular, Descartes found a method for finding tangents to curves, a basic idea behind the derivative in calculus.

Descartes' mother died when he was only 1 year old. At 8 years old, Descartes attended a Jesuit school. In addition to his formal education, Descartes traveled much during his youth and he believed this advanced his education. His father wanted him to study law, which he did at the University of Poitiers, but Descartes was motivated by a dream he had in which he decided to pursue science as a young man.

Descartes applied the scientific method to philosophical thought. He abandoned any philosophical notions that could be doubted, and worked his way up through theories based on an axiomatic system. However, despite accusations of the like, he did not abandon his faith in God because he felt that God's existence could be proven ontologically through *a priori* reasoning. In his famous 1637 philosophical work that greatly influenced modern philosophy, the *Discourse on the Method*, Descartes stated his famous line: "Cogito ergo sum." This means: I think, therefore I am. By this, Descartes meant that although a skeptic may fear that he or she does not really exist and instead is the dream of some dreamer, the fact that one ponders such a notion means that one is thinking. Therefore, in the least, one's own mind exists. This could mean that from a first-person perspective, I, the author, know that I exist. So perhaps, I am the dreamer and everyone else is created by my mind. However, you, as the reader, have the same perspective. So from your perspective, perhaps, it is only you who truly exists. A common joke is that Descartes walked into a restaurant and was asked by the waiter if he would like anything to drink with his dinner. Descartes replied, "I think not," and promptly disappeared.

It was in his great mathematical work, *The Geometry*, also in 1637, that Descartes developed what we call today the Cartesian plane or coordinate plane, an idea that would be essential in the development of calculus. When Descartes was a young boy, he was sickly and was permitted to stay in bed until around noon every day. A legend says that while watching a fly on the ceiling, Descartes wondered how he could describe the fly's location to someone else. He realized that if the ceiling were a coordinate grid, he could give the exact location of the fly. An interesting aspect to his coordinate geometry is the ability to prove theorems from the *Elements* analytically on the coordinate plane. The proofs of Euclid are considered to be synthetic proofs, compared with the analytic proofs possible in Descartes' analytic geometry. It was his work with tangent lines, or lines that touch a curve at only one point, which would greatly influence the development of calculus later in the 17th century.

In the winter of 1649, toward the end of his life, Descartes was brought to Sweden by Queen Christina to be her private tutor. Since Descartes had become accustomed to rising around noon each day, all of his life Descartes had been a late riser. However, the queen wanted her tutoring sessions to be very early in morning. The combination of early rising and the cold Swedish winter may have led to Descartes' early death. Descartes died in February 1650 of pneumonia.

Pierre de Fermat, born in 1601 in southern France, is also credited with developing analytic geometry independently of Descartes. He was a lawyer by trade, but also did work in mathematics as an amateur. His 1636 work, *Introduction to Plane and Solid Loci*, predated Descartes by 1 year. Fermat introduced a method of integration for functions with general powers that would later influence Newton and Leibniz.

Fermat did rigorous work in early probability theory through his correspondence with Blaise Pascal. It was said that a professional gambler had asked Fermat about improving his gambling skills, and this motivated Fermat's work in probability. Along with Pascal, Fermat is considered the "Father of Probability." Recall that Cardano had done work in probability, but not to the same extent as Fermat and Pascal.

Fermat did much of his work in number theory, and his name is attached to one of the most famous mathematics problems in history. Fermat's last theorem states that there are no nontrivial (e.g., $a = 0$, $b = 1$, and $c = 1$) positive integers $a$, $b$, and $c$ that could satisfy the equation $a^n + b^n = c^n$, where $n$ is an integer greater than 2. In other words, the Pythagorean theorem does not work when the exponents are 3 or more. Until 1995 when the theorem was finally proved and published by Princeton University professor Andrew Wiles, its proper name should have been called Fermat's last conjecture because there is no evidence that a proof had been provided. A conjecture is like a hypothesis in science because it is unproved, but a theorem has already been proved. In 1637, Fermat wrote in the margin of his copy of Diophantus' *Arithmetica* that he had found an admirable proof for this but that it was too long to fit in the narrow margin of the page. Many mathematicians doubt if he ever actually found a proof due to the fact that the proof by Wiles is over 150 pages long and it relied on mathematics discovered after Fermat's time. Furthermore, Fermat was known to make mathematical statements without offering proof. The great mathematician Carl Friedrich Gauss said with derision that he was uninterested in this problem because he too could make many conjectures that had not been proven. For over 350 years, mathematicians have attempted and failed to provide a proof, and various organizations have offered prizes for a correct proof. Until Wiles completed his proof, this quest was considered the "Holy Grail" of mathematics.

One notable legend about Fermat's last theorem is how it saved someone's life. Paul Wolfskehl was born in Germany in 1856. The legend says that Wolfskehl was so depressed after he was turned down by a woman whom he loved that he had decided to commit suicide at midnight. After putting his finances in order, he wrote his suicide letter and final will. Because he finished this work before midnight, he decided to pass the time by reading mathematics. He believed he found an error in a proof that would help prove Fermat's last theorem. He worked all night trying to correct it, and when the next morning arrived, he was so satisfied with his work that he decided not to commit suicide after all. He tore up his will and instead decided to offer a substantial monetary prize for anyone who could completely prove Fermat's last theorem. Andrew Wiles received this prize in 1997.

Another legacy Fermat left is known as Fermat's little theorem, which states that for any positive integer $a$ and any prime number $p$, $a^p - a$ is divisible by $p$. This later could be written in the modular arithmetic more fully developed by Leonhard Euler (pronounced "Oiler") in the 18th century and Carl Friedrich Gauss in the 19th century. As was common with Fermat, he did not supply a proof, but claimed again that the proof was too long. Euler, however, was able to provide a proof for this theorem in 1736.

Modular arithmetic is written as follows: $a \equiv b(\text{mod } n)$, which means that $a - b$ is divisible by $n$. It is read, "$a$ is congruent to $b$ modulo $n$" or "$a$ is congruent to $b$ mod $n$." For example, we could have $12 \equiv 6(\text{mod } 3)$ because $12 - 6$, which is 6, is divisible by 3. We could also have $31 \equiv 4(\text{mod } 3)$ because $31 - 4$, which is 27, is also divisible by 3. Therefore, in modular arithmetic, we could write Fermat's little theorem as follows: $a^p \equiv a(\text{mod } p)$. Fermat died in 1665.

Blaise Pascal was born in 1623 in central France; his mother died when he was 3 years old. Pascal was a very bright child and discovered the sum of the interior angles of the triangle on his own. He developed a mechanical calculator, called the Pascaline, between 1642 and 1645, to help his father, who worked as a tax collector, perform computations with less difficulty. About 20 years prior to this invention, the first slide ruler had been developed in England, but the Pascaline was more advanced. Pascal is credited for developing probability theory through his correspondence with Fermat. Similar to the motivations of Fermat, it was a gambler who prompted Pascal to ponder the theory of probability. The work of Fermat and Pascal at first was a direct application to gambling, but was later applied to many other areas of study such as economics and science. The first probability book, *On the Calculations in Games of Chance*, was written in 1657 by Dutch mathematician Christiaan Huygens, born in 1629 and died in 1695. In this work, the idea of expected value is introduced, which informs us how much money we could expect to win or lose by repeatedly participating in a game of chance. It is this idea of expected value that ensures the vast profits of the casinos in Las Vegas, Atlantic City, Monte Carlo, and Macau, among other gambling locations throughout the world. Even if there is only a small expected gain for casinos, staggering profits can be made over the long-term playing of multitudes of customers.

Let us look at a probability example. If we roll two dice, we may want to know what the probability is that we obtain a sum of "6" from both dice. For one die, figuring the probability for obtaining a single "6" would be much easier. We would need to realize there are only six possibilities and that one of these possibilities satisfies the case of obtaining a "6," so the answer would be 1/6. However, for two dice, it is more complicated and we would need to list all possibilities. In Table 12.1, the first number in the parentheses is the number from the first die and the second number in the parentheses is the number from the second die.

**TABLE 12.1  Possibilities for Rolling Two Dice**

| | | | | | |
|---|---|---|---|---|---|
| (1, 1) | (1, 2) | (1, 3) | (1, 4) | (1, 5) | (1, 6) |
| (2, 1) | (2, 2) | (2, 3) | (2, 4) | (2, 5) | (2, 6) |
| (3, 1) | (3, 2) | (3, 3) | (3, 4) | (3, 5) | (3, 6) |
| (4, 1) | (4, 2) | (4, 3) | (4, 4) | (4, 5) | (4, 6) |
| (5, 1) | (5, 2) | (5, 3) | (5, 4) | (5, 5) | (5, 6) |
| (6, 1) | (6, 2) | (6, 3) | (6, 4) | (6, 5) | (6, 6) |

Next, we see that five cases give us the desired sum of "6" along a diagonal starting with (1, 5) and including (2, 4), (3, 3), (4, 2), and (5, 1). Since there are 36 possible cases, the probability of obtaining a sum of 6 on two dice is 5/36.

In 1653, Pascal published his *Treatise on the Arithmetical Triangle*, which presented a fully developed principal of mathematics induction and Pascal's triangle. Recall that Pascal's triangle was well known in earlier centuries in other places throughout the world. Varahamihira in India worked with this triangle in the 6th century. Islamic mathematicians al-Karaji in the 10th century and Khayyam in the 11th century worked with this triangle. Jai Xian in China developed the triangle in the 11th century, but it was called Yang Hui's triangle after the 13th-century Chinese mathematician Yang Hui. Even Tartaglia in 15th-century Italy had worked with it, and it is known as Tartaglia's triangle in Italy. Pascal's triangle (Figure 12.1) helped him to systematically list the binomial coefficients for the binomial theorem, a theorem that would later be generalized by Isaac Newton. A number is generated by adding the numbers directly above, from the left and right, to make the new number. The first 10 rows are as follows. For

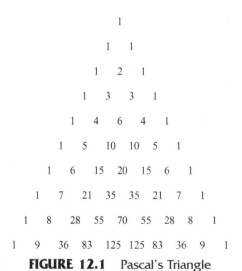

**FIGURE 12.1**   Pascal's Triangle

example, notice that on the fourth row, the first number 3 is found by adding the first number 1 and first number 2 from the third row.

The binomial theorem states that for nonnegative integers $n$, we have $(x + y)^n = C(n, 0)x^{n-0} y^0 + C(n, 1)x^{n-1} y^1 + C(n, 2)x^{n-2} y^2 + \cdots + C(n, n)x^{n-n} y^n$. The coefficients for each term of this expansion are found using the formula for combinations with $n$ greater than or equal to $r$ and both nonnegative integers: $C(n, r) = n!/(n - r)!$. Note that $a!$, read as "$a$ factorial," equals $a \times (a - 1) \times \cdots \times 3 \times 2 \times 1$ (for example, $3! = 3 \times 2 \times 1 = 6$). The notation of using an exclamation mark for factorial comes from German mathematician Christian Kramp. It was created in 1808 due to the surprise of how quickly taking factorials of relatively small nonnegative integers grew, but the idea was known to Indian mathematicians in the 12th century. If using combinations and factorials is confusing, all we need to do is use the values from Pascal's triangle for the coefficients. For example, if we want to expand $(x + 2)^3$, we simply look at the third row, since $n = 3$, to yield the coefficients (really, the fourth row because the first row is considered to be row zero): $1x^3(2)^0 + 3x^{3-1}(2)^1 + 3x^{3-2}(2)^2 + 1x^{3-3}(2)^3$, which simplifies to $x^3 + 6x^2 + 12x + 8$.

In 1654, after he nearly died when his horse carriage almost went over a bridge, Pascal abandoned mathematical pursuits with little exception to devote his life to religion. Near the time of his death, he produced a work called *Pensees*, translated as *Thoughts*, which was not published until after his death and outlined his defense of Christianity. In this work, he attempted to apply his probability theory to the question of God's existence, which is referred to as Pascal's wager. Pascal claimed that if people accept belief in the Christian God and they are correct, then Heaven will be the reward. If they are wrong, nothing will happen because in death there will be nothingness. If they reject the Christian God and they are right, again, nothing will happen. If they are wrong, then they face eternal damnation in Hell. Even if there is a small probability in the Christian God's existence, the expected outcome dictates belief. Pascal's wager has been criticized because he does not account for the claim of competing religions. Biologist Richard Dawkins said in 2006 that

Pascal does not account for competing claims that if God existed, perhaps God would not want to be around people who believed, but rather he would prefer to surround himself with those who questioned. Prominent atheist George H. Smith developed "Smith's wager" in 1976 in response to Pascal. Smith's wager proposed that believing in God's existence for the sake of Pascal's wager is intellectually dishonest and can harm our lives, which means there is something to lose by believing in God's existence. Second, if God punishes people in Hell simply for disbelieving, then God is not just and we cannot rely on an unjust God to reward and punish consistently. Pascal's wager is a valid argument, meaning that it is not possible to accept the premises yet deny the conclusion. However, the soundness of the argument, the truth of the premises, is what comes into question, and the author of this book takes issue with the truth of Pascal's premises. Finally, there would likely be a serious question of faith if anyone believed in God's existence based on the risk analysis of a probability model. It seems disingenuous to base one's religious beliefs on this type of wager. Pascal died before the age of 40 in 1662 of poor health.

Our next 17th-century figure was a Jesuit priest named Marin Mersenne, who was born in 1588 in northwestern France and is known as the "Father of Acoustics" for his work in music. He was a contemporary of Galileo, whom he supported given the troubles Galileo had, and he also knew René Descartes. In mathematics, Mersenne is most famous for his work on prime numbers. Named after him are the Mercenne prime numbers of the form $2^p - 1$, where $p$ is a prime number. The smallest Mercenne prime number is 3 because $2^2 - 1 = 3$. Other examples are 7, 31, and 127, all of which were known to the ancient Greeks. We know, as the ancient Greeks did, that there are infinitely many prime numbers. However, every few years, supercomputers find a new highest known prime number. In recent years, these prime numbers have been Mercenne prime numbers. Since 1997, the Great Internet Mersenne Prime Search Group (GIMPS) has been recruiting volunteers to run software on their computers to find the next Mersenne prime number through the combined efforts of their joined computing power, which has resulted in many Mersenne prime numbers being found.

John Wallis, born in 1616 in southeastern England, introduced the symbol we use for infinity, $\infty$, and the notation for exponents, and understood how to use fractional exponents. He was known for his ability to perform lengthy calculations mentally. Wallis contributed greatly to the development of calculus, especially in the work of integration. His most important work was *Arithmetica Infinitorum*, published in 1656. He died in 1703.

Just as Descartes and Fermat developed analytic geometry independently, Isaac Newton and Gottfried Leibniz also developed calculus around the same time as each other and without consultation. Recall that Archimedes in ancient Greece, Bhaskara II in India, and Islamic mathematicians had developed calculus ideas long before the 17th century in Europe. However, it is widely accepted that Newton and Leibniz unified calculus more comprehensively than any previous mathematician. It is now that we turn to the development of calculus.

Isaac Newton, like Archimedes, is considered one of the greatest mathematicians of all time. His contributions to mathematics and physics were unparalleled. Even to this day, there are very few who could compete with Newton. Along with Gottfried Leibniz, Newton received credit for developing differential and integral calculus, and, like Leibniz, is considered to be the "Father of Calculus." Newton developed Newtonian physics, often called classical mechanics. Although still valid today, it was not superseded until Albert Einstein's theory of relativity in the early 20th century.

Newton was born in England on Christmas Day in 1642, using the Julian calendar that England still used at that time, but his birth was in January 1643 by the modern Gregorian calendar. His father died a few months before he was born. Newton despised his stepfather. In his early years, Newton did not perform to his academic abilities, but later showed great promise. In 1661, he entered Cambridge University's Trinity College. Interestingly, despite his mother's wealth, Newton entered the university in a work–study capacity through which he served other students in order to help pay for his tuition. Cambridge operated under under Aristotle's philosophy, but Newton was more interested in the work of Copernicus, Kepler, Galileo, and Descartes. At Cambridge, Newton studied under English mathematicians Isaac Barrow, who was born in 1630 in London and was appointed as the first Lucasian Chair of Mathematics in 1663. Barrow had great influence on Newton's development of calculus, and is often given credit for the early stages of its development. He worked on finding the tangents to points on curves and the area under curves, which are the central components of differential and integral calculus. Barrow worked on the fundamental theorem of calculus, which demonstrates the relationship between the derivative and the integral, and allows us to find the definite integral. Newton more fully developed the fundamental theorem of calculus. It should be noted that the fundamental theorem of algebra, which involves the number of roots for a polynomial, was only recently explicated.

Newton graduated from Cambridge in 1665, but shortly after graduation the university shut down for 2 years during the Great Plague that affected England during that time. It was during these 2 years at home that Newton was extremely productive and developed his ideas of calculus in addition to his theories on optics and gravitation. The legend about gravitation says that an apple hit Newton on the head, which inspired his theory. However, it is more likely that Newton simply observed an apple fall without hitting him on the head. Newton returned to Cambridge in 1667 and Barrow recommended Newton replace him as Lucasian Professor of Mathematics in 1669.

In 1672, Newton published his first work, which was well received except by a few, including fellow Royal Society member Robert Hooke. Newton and Hooke developed a rivalry over the years, further exacerbated by when Hooke accused Newton of stealing his work on optics in 1675. Their rivalry manifested in the form of polite letter correspondence between the two of them with some hidden insult. In a letter to Hooke, Newton alluded to a famous line from the 12th-century French philosopher Bernard of Chartres. Newton complemented Descartes because his work informed Newton's own, but then added that if Newton had seen further than others, it is because he has stood on the shoulders of giants, meaning that Newton was relying on the great works that came before him. The implication is that the foundation for Newton's work has nothing to do with Hooke, who is often believed to be short, but was actually hunch-backed instead. This quote was incorporated into a book title by a recent Lucasian Professor of Mathematics, Stephen Hawking. Today, the British £2 coin has the words, "Standing on the Shoulders of Giants" on its edge. Incidentally, Newton has appeared on the former British £1 note as well.

Newton hesitated to publish his calculus because of his fear of criticism and the difficulty he confronted when searching for a publisher. However, in 1684 Gottfried Leibniz published his work in calculus. Although both had developed their work independently, it is likely that Newton developed calculus first, but Leibniz published calculus first. Generally, both are given credit for the discovery of calculus. In 1687, Newton published his *Philosophiae Naturalis Principia Mathematica*, or *Mathematical Principles of Natural Philosophy*, often called *Principia Mathematica*. It is in this work that Newton outlined his calculus and used the book to present his physics theory. *Principia Mathematica* provided Newton's theories on motion and gravitation that laid the foundation for Newtonian physics or classical mechanics, and it is considered to be the most important scientific work ever published.

Newton's three laws of motion are as follows:

1. Every body stays at rest or in constant velocity unless acted on by an outside force, which is inertia.
2. A body's mass multiplied by its acceleration equals its force: $F = ma$.
3. A force acts on two bodies equally in opposite and equal collinear directions, which means that every action is followed by a reaction in opposite direction but equal in magnitude. In Latin this is "actio et reactio" or "action and reaction."

Newton's law of universal gravitation states that bodies attract each other with a force directly proportional to their masses, but inversely proportional to the squared distance between them.

In 1689, Newton became a member of Parliament, and in 1699, he was in control of England's Mint. In 1693, Newton suffered a nervous breakdown and afterward focused mainly on his work in government and rarely practiced mathematics or science. Newton became president of the Royal Society in 1703 after resigning from his Lucasian Chair of Mathematics position at Cambridge University. Two years later in 1705, Newton was knighted by Queen Anne. In 1715, the Royal Society determined that Newton, and not Leibniz, was the founder of calculus. This was viewed harshly by many because Newton was the one making this decision as president of the Royal Society. He spent much of the remainder of his life in despair over the debate with Leibniz regarding the priority of the calculus. Toward his final years of life, it is possible that he became a Pythagorean, meaning he possibly became a vegetarian. He died in 1727 and was buried in Westminster Abbey in London.

Gottfried Leibniz, the "Father of Calculus" along with Newton, was born in 1646 in Leipzig in Saxony, which is now in eastern Germany. Due to his great accomplishments in many fields including philosophy, history, language, engineering, science, logic, and mathematics, Leibniz is considered to be a polymath. Like Newton, Leibniz grew up without his biological father because Leibniz's father died when he was 6 years old. Although Newton's father owned land and was wealthy, he was uneducated. Liebniz's father, however, had been a professor of philosophy at the University of Leipzig, which meant that Gottfried inherited his father's library. During his childhood, Liebniz had a propensity toward languages and learned Latin and Greek very quickly. Liebniz later studied at his father's university and went on to study law. Like Newton, Liebniz completed a degree in 1665. In 1666, Liebniz completed his first book, which was on combinatorics and called *On the Art of Combinations*.

In 1672, Leibniz went to Paris to discuss Franco-German politics with the French government, but he began studying mathematics in

Paris under Dutch mathematician Christiaan Huygens. It was in 1672 that Leibniz began to work on his Step Reckoner, which was an improvement of the mechanical calculator, like the Pascaline before him. It was the first mechanical calculator that could perform the four basic operations of addition, subtraction, multiplication, and division. However, due to technical limitations during that time, it did not function properly. The Step Reckoner would be succeeded by Charles Baggage's Difference Engine in the early to mid-19th century, which is considered to be the first mechanical computer.

It was during Leibniz's time in Paris that he began his independent development of calculus, over 5 years after Newton. In 1675 and 1676, Leibniz developed his calculus notation that would become much more influential to modern calculus and is for the most part the notation we use today. Even outside calculus specifically, Leibniz popularized the equals sign for equality and the $\times$ for multiplication. While Newton used dot notation for the derivative by placing a dot over $x$ ($\dot{x}$), Leibniz represented the derivative as $dy/dx$. The $d$ is from the Latin for differentiate, which is "differentia." He used $\int f(x) \, dx$ for the integral with the integral "$\int$" symbol used because he believed that the area under a curve, found through the integral, could be considered a summation, or in Latin "summa," of infinitesimals. This idea is similar to the indivisibles of Archimedes. Leibniz also gave us the terms "differentiation" and "integration." French-Italian mathematician Joseph Louis Lagrange, born in 1736 in northwestern Italy and died in 1813, used the "prime" notation, which is still used today: $f'(x)$ Although there were many other mathematical terminologies named after Lagrange, he may be best known to calculus students for Lagrange multipliers. Lagrange took calculus to a new level of rigor and could be considered the first to engage in true analysis.

From the time of Newton and Leibniz onward, British mathematicians used Newton's notation, while Continental European mathematicians used Leibniz's notation. This gave the Continental mathematicians an advantage over the British mathematicians and led to more advancement in Continental Europe. The result can be found in the incredible advancements made in mathematics in

France and then Germany, along with mathematicians in the countries nearby such as Switzerland, among others, in the 18th and 19th centuries. Eventually, British mathematicians would adopt Leibniz's notation.

In 1676, Leibniz moved from Paris to Hanover to be a librarian under the Duke of Hanover. On the way there, he visited the Royal Society, and during his second visit, he may have seen some of the work Newton had done in calculus, which is called the method of fluxions and fluents. "Fluxions" was Newton's word for differentiation and "fluents" was his word for integration. This led to accusations later that Leibniz had stolen Newton's work.

In 1682, Leibniz helped found the first scientific journal in Germany called *Acta Eruditorum*, or *Acts of the Scholars*. In 1684, Leibniz published his first work on calculus in the journal called *New Method for Maximums and Minimums*, 3 years prior to Newton's *Principia Mathematica*. In 1686, 1 year before Newton's publication, Leibniz published another article in the journal presenting his ideas more fully on integration. In 1671, Newton had written a work, *Method of Fluxions*, but it would not be published until 1736. For a while, the issue of the foundation of calculus was fairly civil between Newton and Leibniz. The great debate on the priority of calculus fully developed in 1711 when Leibniz was directly accused of plagiarism by a supporter of Newton. The debate had supporters of Newton and Leibniz on both sides. Like Newton, Leibniz was very distraught by the calculus debate in his later years.

Toward the end of his life, Leibniz wrote on theological issues, including addressing the problem of evil. Leibniz claimed that the universe was the best possible universe because anything else would be perfect and would not be any different from God. Like Pascal's argument, this assertion is open to criticism because although Leibniz claimed that if certain evils in the world were removed, such as natural disasters, the laws of science would be disrupted, he does not account for the concept that an omnipotent deity could design such a world that, although not perfect, would have less suffering than our own. Leibniz died in 1716.

Let us now look at some calculus. Differential calculus is based on the idea of the derivative, which is the rate of change at a given point on the curve. In other words, we could find the slope of the tangent line at the point on the curve if we want to find the derivative at that point. In a curve, this is changing depending on where on the curve we want to observe. We are looking for a way to find a function for the derivative. To find the derivative, we can essentially draw a secant line (Figure 12.2), a line that passes the curve at two points through the curve, and keep making this secant line approach a tangent line (Figure 12.3), which is a line that passes through one point on the curve. As this difference gets smaller, we are approaching the derivative.

First, we need to remember that the slope of a line is the rise divided by the run, which is $(y_2 - y_1)/(x_2 - x_1)$ or, in function notation,

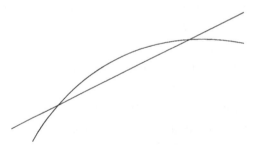

**FIGURE 12.2**   Secant Line to a Curve

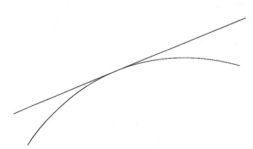

**FIGURE 12.3**   Tangent Line to a Curve

$(f(x_2) - f(x_1))/(x_2 - x_1)$. For the secant line with the points getting closer and closer, we can find the slope of the secant line, which is the idea of derivative:

$$f'(x) = \frac{f(x+h) - f(x)}{x+h-x} = \frac{f(x+h) - f(x)}{h}$$

Since we want the distance to be smaller and smaller, we are essentially saying that $h$ is approaching zero, and this is our definition of the derivative. Finding the derivative of $f(x) = x^3$ as $h$ approaches zero, we yield

$$f'(x) = \frac{(x+h)^3 - x^3}{h} = \frac{x^3 + 2x^2h + xh^2 + x^2h + 2xh^2 + h^3 - x^3}{h}$$
$$= \frac{h(2x^2 + xh + x^2 + 2xh + h^2)}{h}$$

After we cancel the $h$'s in the numerator and denominator and let $h$ approach zero, we get:

$$f'(x) = 2x^2 + xh + x^2 + 2xh + h^2 = 2x^2 + x^2 = 3x^2$$

To find the derivative of a general power function, that is, $f(x) = x^n$, we can write:

$$f'(x) = \frac{(x+h)^n - (x)^n}{h}$$

as $h$ approaches zero. We can use the binomial theorem to expand $(x + h)^n = C(n, 0)x^nh^0 + C(n, 1)x^{n-1}h^1 + C(n, 2)x^{n-2}h^2 + \cdots + C(n, n)x^0h^n$. Since $C(n, 0) = 1$, because $(n!/(n - 0)! = 1)$ and $h^0 = 1$, we can see when we subtract $x^n$ from the $(x + h)^n$ expansion, we get the first term of the expansion canceled. With the rest of our expansion, we factor out a common $h$ to yield

$$f'(x) = \frac{h[C(n, 1)x^{n-1} + C(n, 2)x^{n-2}h + \cdots + C(n, n)x^0h^{n-1}]}{h}.$$

The $h$ in the numerator cancels with the $h$ in the denominator, and all of the other $h$'s approach zero. Since $C(n, 1) = n$, because

$n!/(n-1)! = n$, we are left with $f'(x) = nx^{n-1}$, which was developed by Leibniz and is a very recognizable formula used in modern calculus textbooks. This makes life much easier because if we want to find the derivative of $x^3$, we only need to place a 3 in front, and subtract 1 from 3 in the exponent to yield $f'(x) = 3x^2$. This is much easier than using the definition of the derivative in the previous paragraph.

The derivative can make modeling real-life phenomena much easier. For example, Newton was very interested in the calculus of motion. This differs from Leibniz's approach, which focused more on differences and summations. If we consider a distance function that gives us the distance of an object, we can take the derivative to get the distance over time, or velocity. Taking the second derivative gives us the velocity over time, or acceleration.

The integral is the reverse of the derivative, similar to the relationships between addition and subtraction or multiplication and division. So we can write the integral of a function $x^n$ as follows:

$$\int x^n dx = \frac{x^{n+1}}{n+1} + C$$

where $C$ is a constant. Since the derivative and integral are reverse operations, we can take the derivative of our result to see if we get back to $x^n$. Checking our work will demonstrate that this works. If we integrate over a closed boundary on the curve, we shall find the area under that curve.

While Newton found the integration by parts and substitution that we use today, Leibniz found the commonly used product and quotient rules of differentiation. Leibniz's notation made the chain rule easy to understand. Leibniz was interested in finding relative maximums, minimums, and inflection points using calculus. The works of both Newton and Leibniz make up much of the calculus textbooks present today.

Our next chapter moves on to the 18th century, a time of great change in the world. The 18th century saw the start of the Age of Enlightenment, which may have began as far back as 1637 with Descartes' *Discourse on the Method* or 1687 with Newton's *Principia Mathematica*. In the next chapter, we shall also encounter some great mathematicians, including a great family of mathematicians, the Bernoulli family, and perhaps the most prolific mathematician to have ever lived, Leonhard Euler.

# CHAPTER 13

# EUROPEAN MATHEMATICS: THE 18TH CENTURY

The Age of Enlightenment, also called the Age of Reason, started around the beginning of the 18th century, and it was the natural product of the Scientific Revolution that used reasoning to find truth separate from religious dogma. During the 18th century, the 13 British colonies in North America formed a new democracy, the United States of America, a result of the American Revolution in 1776. The French overthrew the monarch in 1789 and executed Louis XVI during the French Revolution, which is generally considered the end of the Enlightenment Period. Both societies were influenced by the notion of citizenship rights as proposed by philosophers John Locke and Jean-Jacques Rousseau. Other notable philosophers such as Voltaire, David Hume, and Immanuel Kant had significant influence during this period. It was during the 18th century that the Classical Period in music took hold and gave birth to the great composers such as Wolfgang Amadeus Mozart and Ludwig van Beethoven.

In 1707, the Kingdoms of England and Scotland united to form the Kingdom of Great Britain. In the mid-18th century, the Industrial

*The Development of Mathematics throughout the Centuries: A Brief History in a Cultural Context*, First Edition. Brian R. Evans.
© 2014 John Wiley & Sons, Inc. Published 2014 by John Wiley & Sons, Inc.

Revolution began in Great Britain and led the nation to the dominant position in the world, which helped build the powerful British Empire, which would not reach its apex until the following century. The Industrial Revolution can be traced back to the adoption of the steam engine in Great Britain, by which time machines began to replace human and animal power in manufacturing and transportation. The Industrial Revolution spread around the world from Europe to the United States, and eventually much of the planet. It brought great prosperity to many, but led to much misery in the form of pollution and mechanized sweat shops. The Industrial Revolution is perhaps the most significant event in human history since humans began to shift from nomadic hunter–gatherers to an agricultural-based society. It was a significant factor in creating the Great Divergence that led to immense wealth in Western Europe and North America while much of the rest of the world remained in poverty. In the present early 21st century, we are experiencing another type of event with the massive growth of technology and the ease of the spread of knowledge through the Internet.

In the 18th century, British mathematicians studied and expanded on Newton's calculus, and Continental European mathematicians studied and expanded on Leibniz's calculus. Likely attributed to the more convenient notation used by Leibniz, there were greater accomplishments on the Continent than in Britain during this time. However, there were still accomplishments in Britain. Brook Taylor, born 1685 near London, developed the Taylor series and Taylor's theorem, which was also developed in part by Newton, Leibniz, and Johann Bernoulli. Ideas of the Taylor series were known in the 14th century by Indian mathematician, Madhava, who was presented in Chapter 8. At that time, Taylor was engaged in his own priority dispute about another topic with Johann Bernoulli, a supporter of Leibniz. Colin Maclaurin, born in Scotland in 1698, received a degree from the University of Glasgow at 14 years old; his thesis expanded on Newton's gravitation ideas. He developed a special case of the Taylor series, which would be called the Maclaurin series.

We now turn our attention to supporters of Leibniz. In particular, we look to the family of mathematicians with the surname Bernoulli. The Bernoulli family of mathematicians originated in Basel, which is located in Switzerland at the border of France and Germany. The first two Bernoullis are the brothers Jacob, born 1654, and Johann, 1667. Their father, Nicolaus, intended for Jacob to study theology and Johann to study business, in order to manage the family's spice trade. However, both brothers became interested in mathematics and studied at the University of Basel. They collaborated for a while but eventually became rivals. Jacob, who began the tradition of mathematics in the family, learned calculus from the work of Leibniz and did his own work in differential equations. His greatest publication is *The Art of Conjecture*, not published until after his death, in which Jacob contributed greatly to the development of probability theory, including the law of large numbers, which states that over the course of many probability trials, the average of the outcomes will approach the theoretical probability. An example of this is flipping a coin enough times until we observe about 50% heads and 50% tails. Jacob is the Bernoulli who is credited with devising the Bernoulli trial, which is a probability experiment in which the results could be success or failure. An example of this can again be given by flipping a coin. Over $n$ number of trials, the probability can be stated for a fixed number of successes as $P(x) = C(n, x)p^{n-x} q^x$, where $P(x)$ is the probably of $x$ number of success, $p$ is the probability of a single success, and $q$ is the probably of a single failure. This will be familiar to the reader from the development of the binomial theorem in the last chapter, in which if we expanded this fully, we would have the binomial distribution.

Johann is best known for his support of Liebniz against Newton in the calculus priority debate. In 1713, Johann solved problems that could be solved using the calculus of Liebniz but not with the calculus of Newton. Johann taught calculus to Guillaume de L'Hôpital, a French mathematician born in 1661 and died in 1704, who is the mathematician responsible for writing the first calculus book and has L'Hôpital's rule named after him. However, it was likely Johann who had developed the rule. L'Hôpital's rule states in its simplest form that if the limit of the ratio of two functions approaches zero or infinity, then the ratio of the derivative of those functions equals the limit of the original ratio of functions. Johann also taught mathematics to Euler, who would become the greatest mathematician of the century. Jacob died in 1705 and Johann died in 1748.

The next generation of Bernoullis was the nephew of Jacob and Johann, Nicolaus I, who was born in 1687, and Johann's three sons, who were Nicolaus II, Daniel, and Johann II. They were born in 1695, 1700, and 1710, respectively. Nicolaus I learned mathematics from his uncles, and especially from Jacob. In 1713, Nicolaus I published Jacob's *The Art of Conjecture*, and in 1716, he occupied the chair formally held by Galileo at the University of Padua. He died in 1759. Nicolaus II began studying mathematics at a young age and helped his father write letters about the priority of calculus debate between Newton and Leibniz, as well as the debate with Taylor. He died at 31 in 1726 due to a fever. Daniel was not born in Basel, like most of the family, but rather in The Netherlands. However, at a young age, he went to Basel with his family and also studied mathematics early in life, like his older brother. He spent time at the Imperial Russian Academy of Sciences as chair of mathematics, in which Euler would succeed him in 1733. Daniel did significant work in probability and statistics, and he died in 1782. Johann II won several mathematics competitions and his two sons were the last of the Bernoulli family mathematicians. Johann II died in 1790. His son, Johann III, was born in 1744 and gained the chair position at the Berlin Academy at only 19 years old. Jacob II was born in 1759 but died at only 29 years old in St. Petersburg, Russia, in 1789. Johann III lived until 1807.

The most influential mathematician of the 18th century was Leonhard Euler (pronounced "Oiler"). He was born in 1707 in Basel, like the Bernoulli family. Euler was one of the most prolific mathematicians to have ever lived and is considered one of the greatest mathematicians of all time. Euler worked in many areas, including algebra, geometry, trigonometry, calculus, number theory, analysis, cartography, and graph theory. No brief summary of Euler's work, as is attempted by this book, can truly do it justice.

Euler earned a Master's degree in philosophy in 1723 from the University of Basel in which his final thesis involved the comparison of the philosophies of Descartes and Newton. Euler's father had been friends and associates with the Bernoulli family. It was Johann Bernoulli who had recognized Euler's potential for mathematics and who subsequently taught Euler mathematics. Euler continued his education in mathematics and completed his education in 1726.

Euler is responsible for giving us much of the mathematical notation we used today. At around 1726, Euler began using $e$, today called Euler's number, to represent the irrational number approximated by 2.7182818285. Although Euler had begun using the letter "$e$" for this number, he did not discover this irrational number. However, he did establish that it was, in fact, an irrational number. Although John Napier alluded to this number in 1618, Jacob Bernoulli is given the credit for finding the value of $e$ by contemplating what number the following expression approaches as $n$ goes to infinity: $(1 + 1/n)^n$. Euler's number is very commonly found in many areas of mathematics and is often used for the exponential function $e^x$, as well as the common usage of $e$ as the base for the natural logarithm. An interesting aspect of the function $e^x$ is that it is the derivative and integral of itself. It also models real-world exponential growth and can be used for continuous compounding. If we take the compound interest formula $A = P(1 + r/n)^{nt}$, where $A$ is the new amount, $P$ is the principal, $r$ is the rate, $n$ is the compound period, and $t$ is the time, we can find a formula in which $n$ approaches infinity for infinite compounding. If we allow $x = n/r$, we can write our formula as $A = P(1 + 1/x)^{xrt}$, which equals $P[(1 + 1/x)^x]^{rt}$. We realize that if $n$ approaches infinity, $x$ does as well. This means the quantity inside the brackets becomes $e$ and our new formula is $A = Pe^{rt}$, a formula that may appear familiar to the reader. Thus, if we have \$100 at 5% interest compounded continuously for 2 years, we have $A = 100e^{0.05(2)}$, which is about \$110.52.

Although Euler wanted to teach at the University of Basel, he ended up taking a position at the Imperial Russian Academy of Sciences in 1727, an academy established by Peter the Great. He was afforded this opportunity in St. Petersburg, Russia, through help from the Bernoullis. In 1733, Euler succeeded Daniel Bernoulli as chair of mathematics, but left Russia in 1741 due to the increasing hostility the Russian government against foreign professors. During his time in Russia, Euler established a family by getting married and fathering five children.

A mathematical function is the relationship between two variables such that for every independent variable, there is only one dependent variable. For example, $y = x + 1$, or in function notation $f(x) = x + 1$, can be seen to have every $x$ yielding one unique $y$. We can think of a function machine, such as a vending machine. When a person presses "E4," for example, he or she gets potato chips every time. The person would be very surprised if pretzels came out if the "E4" code is under the potato chips section. For every input, there is one unique output. This gives us the power of predictability, and is an important aspect of mathematics. Although Euler did not discover the concept of functions (it was discovered instead by Leibniz in 1673), in 1734 Euler began using the familiar function notation $f(x)$, read as "$f$ of $x$." It was Euler's work with functions, along with Leibniz's version of calculus, which led to mathematical analysis.

Euler's development of graph theory began with his 1735 Seven Bridges of Königsberg solution, which later led to topology. Due to this discovery, Euler was considered to be the "Father of Graph Theory." The Seven Bridges of Königsberg problem involved the islands and bridges in the city of Königsberg, Prussia, which is now Kaliningrad, Russia. Königsberg was separated by the Pregel River with two islands in the river. The islands were connected to the mainland, as can be seen in Figure 13.1.

The problem was to find a way to walk around the city by crossing every bridge only once. Euler showed that this could not be done. The land masses could be represented as vertices and the bridges could be represented as edges to create a graph (see Figure 13.2). One

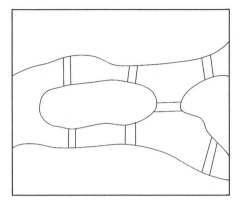

**FIGURE 13.1** Königsberg Bridges

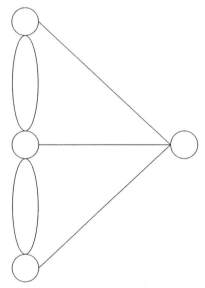

**FIGURE 13.2** Königsberg Graph

must cross a bridge to get to a land mass, and then leave that land mass by another bridge. However, in Königsberg, each land mass had three bridges connecting to it, and the central island had five bridges. Euler showed that it was impossible to cross all bridges only once since there were more than two land masses connected by an odd number of bridges. During World War II, the Allies bombed the city and destroyed two of the bridges connecting the island. As a result, only five bridges remain, and the two land masses have only two bridges. If a person begins the tour on one of the islands, it is possible to complete a circuit by crossing every bridge only once.

Euler also developed a formula relating the vertices, edges, and faces of the five Platonic solids by observing that for each solid, the number of vertices plus the number of faces minus the number of edges always equaled 2. In other words, $V + F - E = 2$, where $V$ is the number of vertices, $F$ is the number of faces, and $E$ is the number of edges. This means that if two of the three variables are known, number of vertices, faces, and edges, the final variable can be found.

As referenced in the last chapter, Euler had developed modular arithmetic and proved Fermat's little theorem in 1736. Euler based much of his work in number theory on the work presented by Fermat. Recall that Fermat often made conjectures, but failed to provide proofs. Euler proved some of the conjectures by Fermat, but he would also disprove other claims made by Fermat.

In 1735, Euler began to loose sight in his right eye, which he blamed on eyestrain due to his work in cartography. He developed a cataract in his right eye in 1766, leaving him nearly blind. Despite his limitations in sight, he continued to be productive due to his remarkable memory. After leaving St. Petersburg in 1741, Euler was given a position at the Berlin Academy by Frederick the Great. It was here that Euler would be very productive and write over 380 mathematical articles.

In 1748, Euler developed the most beautiful formula ever known in mathematics. This formula is known as Euler's identity: $e^{i\pi} + 1 = 0$. The beauty in this identity is that it brings together five of the most important numbers in mathematics, including Euler's number $e$, the imaginary number $i$, the irrational number $\pi$, 1, and 0. As stated earlier, Euler gave us the notation for $e$, and it was also Euler who popularized the notation for $i$ and $\pi$. Euler also gave us the notation for summations by using sigma ($\Sigma$), the change in $x$ notation by using "delta $x$" ($\Delta x$), and the modern trigonometric notation for sine and cosine functions. Euler's textbooks in calculus—*Introduction to the Analysis of the Infinite* in 1748, *Method of Differential Calculus* in 1755, and *Methods of Integral Calculus* in 1768—proved to be extremely influential. In 1765, his *Elements of Algebra* presented algebra in a manner in which we may recognize it today. It was Euler who gave calculus an emphasis on functions rather than on geometry.

Euler stayed in Berlin until 1766; when he returned to St. Petersburg, Catherine the Great was then in power, and the hostility toward foreigners had ceased. In 1769, Euler discovered a method of double integration. He spent the rest of his life in St. Petersburg until his death of brain hemorrhage in 1783. On his death bed, his final words were the simple words that he was dying.

There is an interesting legend about Euler that is very unlikely to be true. French philosopher and atheist Denis Diderot, born 1713 and died in 1784, visited Russia. Catherine the Great was upset by his atheism, and she requested help from Euler to prove God's existence to Diderot. Euler approached Diderot in Catherine's court and said that since $(a + b^n)/n$, God exists, and then demanded a response. The legend says that Diderot did not understand mathematics and could not respond. However, Diderot was known to be quite proficient in mathematics, which puts serious doubt on the accuracy of this legend. If true, Euler is even less convincing in his theological argument than Pascal and Leibniz were in their own arguments. Interestingly, we have seen so far that many of the mathematicians from this period were very religious, yet today in the United States, only 7% of members of the National Academy of Science claim belief in a personal god.

We have seen some of the development of probability theory from Cardano in Chapter 11, Fermat and Pascal in Chapter 12, and Jacob Bernoulli in this chapter. Abraham De Moivre was born in 1667 near Paris. He was imprisoned for his Protestant beliefs in Catholic France, so on his release in 1688, he fled to London. He often traveled from one student's house to another's house to tutor and was so fascinated by Newton's *Principia Mathematica* that he tore pages out of the book so he could always have some pages with him to study as he traveled house to house. His book on probability was the 1718 publication *The Doctrine of Chances*, which was highly valued by professional gamblers at that time. It introduced the concept of the normal curve and central limit theorem, which would prove extremely valuable in later probability and statistics. The central limit theorem states that the distribution of a large enough number of random variables will be normally distributed. A major concern for De Moivre was the concept of annuities. An annuity is a certain amount of money paid over a given amount of unknown time. For example, in retirement, one may have an annuity that pays until death. The longer the duration of one's life, the more one benefits; and the shorter one lives, the less one benefits. Some people put more into the annuity in their lifetimes than they will receive, while others put less in than they will receive. An insurance company would need to figure out sufficient probabilities to determine the value of such annuities, a problem interesting for De Moivre. De Moivre died in 1754.

Thomas Bayes was an English minister born in 1702 in London who studied theology and logic at the University of Edinburgh. He published two works; one was religious and the other was on probability theory. His major work on probability theory was *Essay towards Solving a Problem in the Doctrine of Chances*, which was not published until 1763, 2 years after his death. It is in this work that Bayes introduced a theorem based on conditional probability, that is, the well-known Bayes' theorem. This theorem states that the probability of event $A$ occurring, given that $B$ has already occurred, is the probably that event $B$ occurs, given $A$ has already occurred, multiplied by the probability of $A$, and all divided by the probability of $B$. Symbolically, this is $P(A|B) = P(B|A)P(A)/P(B)$, reading $P(A|B)$ as "the probability of $A$ given $B$ has occurred." An example of this illustrated by the following. Suppose one has five red marbles and five blue marbles in a bag. If one had already drawn a red marble, the individual wants to find the probability that the next draw is a blue marble. One's answer is based on conditional information. Bayes' theorem is derived from basic conditional probability: $P(A|B) = P(A \cap B)/P(B)$, where $P(A \cap B)$ is read "$A$ intersects $B$." Since we also have $P(B|A) = P(A \cap B)/P(A)$, we can solve for $P(A \cap B)$ in both equations to get $P(A \mid B)P(B) = P(B \mid A)P(A)$. Solving for $P(A \mid B)$ yields Bayes' theorem.

Pierre Simon Laplace was a French mathematician and astronomer born in 1749 in Normandy in northwestern France, and he has been called the "Isaac Newton of France." There has been some debate regarding the beginnings of Laplace. Some have said that he was from a poor family, but now it is generally believed that Laplace was from a moderately wealthy family. Laplace studied theology at the University of Caen in Normandy, but he did not finish his degree and instead became interested in mathematics. He next went to Paris where he quickly demonstrated his mathematical genius to mathematician Jean le Rond d'Alembert, who secured a position for Laplace at École Militaire. Laplace left Paris in 1793 during the Reign of Terror during the French Revolution and avoided the guillotine, unlike some of his colleagues.

Laplace is remembered for his development of probability theory from Bayes' interpretations and for his contributions to the development of statistics. Laplace used calculus in his probability models to derive many of the ideas proposed by Bayes. He developed the Laplace transform and a method for least squares regression. His 1812 book, *Theorie Analytique des Probabilites*, was originally dedicated to Napoleon, whom he had examined in school when Napoleon was 16 years old. Laplace would later briefly serve under Napoleon as minister of the interior. It turned out that Laplace was a much better mathematician than a government worker. He later removed his dedication to Napoleon and campaigned against Napoleon in 1814. An interesting anecdote exists about Napoleon and Laplace. Napoleon had noticed that Laplace did not mention God in one of his books, so Laplace replied that he had no need for that hypothesis.

Laplace is remembered for his concept of "Laplace's demon" or sometimes called "Laplace's superman." The idea is that the universe is deterministic, which means that all events are bound by causation. In other words, the universe is like a giant complicated set of dominoes. The demon is the omniscient mind that would be able to calculate all forces in the universe and predict with certainty all events. In other words, probability only exists because we do not know everything that one would need to know. Instead of a 50% probability of a coin landing heads, it is either 100% heads or 0% heads if we could know the magnitudes of all of the forces involved as well as all of the angles involved in the flip. This idea has been criticized because sometimes knowing something affects the scenario, such as Heisenberg's uncertainty principle, which states that it is not possible to know both the velocity and position of an electron since measuring one affects the other. Perhaps Laplace's demon could be reconciled with the idea that there could not exist an omniscient being who could know the outcomes, but rather the outcomes are purely deterministic alone. Or perhaps one could consider a hypothetical omniscient mind that does not get involved with the perceived probabilistic scenario taking place. Laplace died in 1827.

We now turn our attention to a female Italian mathematician named Maria Gaetana Agnesi, who was born in Milan in 1718. Agnesi came from a wealthy family that enabled her to have the tutors she needed to learn mathematics. She was a child prodigy who had mastered many languages at an early age. Agnesi published a book on calculus, *Instituzioni Analitiche Ad Uso Della Gioventu Italiana*, in 1748. In this book, Agnesi discussed a mathematical curve that has come to be known as the "witch of Agnesi." This curve appeared in the book in Italian as the "curve of Agnesi." The word "curve" was mistranslated as "witch." This curve was first addressed by Fermat, but it is interesting that the world "witch" was retained given that this curve is named after a female mathematician. Agnesi was awarded the position of chair of mathematics at the University of Bologna in 1750, but she declined the offer because at that point in her life, she was devoted to a life of religion and charity. Agnesi died in 1799.

The 18th century saw interest in studying geometry without Euclid's controversial parallel postulate. As we saw in Chapter 9, Islamic mathematicians such as Alhazen and Khayyam in the 11th and 12th centuries were troubled by the parallel postulate. In Europe, Italian priest and mathematician Giovanni Saccheri, born in 1667 in northwestern Italy, created a work called *Euclid Freed of Every Flaw*, which was published in the year of his death in 1733 and possibly grounded Khayyam's work. Saccheri presented the Saccheri quadrilateral, a concept also known to Khayyam. The Saccheri–Khayyam quadrilateral consisted of two right base angles with equal sides rising from the base. Saccheri wished to show that the angles opposite these right angles were also right, but he did not want to do this using the parallel postulate. Rather, he wanted to prove the parallel postulate. His work would eventually develop into hyperbolic geometry.

Johann Lambert, Swiss mathematician born in 1728 and died in 1777, who is probably most famous for his rigorous proof of the irrationality of $\pi$, would attempt to improve on Saccheri's work in geometry. In 1766, Lambert wrote *Theory of Parallel Lines*, which derived non-Euclidean results by assuming that the parallel postulate was false. Non-Euclidean geometry would develop more fully in the 19th century.

We see that the 18th century added to many of the accomplishments of the great 17th century. We shall continue to see more significant advancement in mathematics in the 19th century, as well as other great accomplishments with the Industrial Revolution in full swing. We shall also examine perhaps the greatest mathematician to have ever lived, Carl Friedrich Gauss.

# CHAPTER 14

# EUROPEAN MATHEMATICS: THE 19TH CENTURY

During the 19th century, great shifts in political power took place. The Napoleonic Wars exhausted much of Europe and led to the British Empire rising to be the dominant power in Europe and in the world in the 19th century. Following the French Revolution, Napoleon Bonaparte rose to power and succeeded in conquering most of Europe. In 1801, the Kingdoms of Great Britain and Ireland united to become the United Kingdom of Great Britain and Ireland, and during much of the 19th century, the British Empire was ruled by Queen Victoria. The Battle of Trafalgar in 1805 signified the victory of the British Navy against the French and Spanish navies, which solidified British dominance of the seas. Napoleon suffered great losses in Russia in 1812, which proved fatal to his conquests and led to his defeat by the British at Waterloo in 1815. The Napoleonic Wars resulted in the dismantling of the Holy Roman Empire by France and the weakening of the Spanish Empire while Spain was occupied by France. In 1871, both Germany and Italy became unified states. In the Americas, the acquisition of the Louisiana Purchase from Napoleon in 1803 resulted in

*The Development of Mathematics throughout the Centuries: A Brief History in a Cultural Context*, First Edition. Brian R. Evans.
© 2014 John Wiley & Sons, Inc. Published 2014 by John Wiley & Sons, Inc.

the United States nearly doubling in size. During the 19th century, the United States grew to be a world power and acquired the rest of the land that make up its present-day states from Great Britain, Spain, Mexico, Russia, and the former republics of Texas and Hawaii. By the late 19th century, the United States had replaced the United Kingdom in having the world's most dominant economy.

In the 19th century, great advancements were made in technology, science, transportation, and medicine. The Industrial Revolution progressed from Great Britain throughout much of Europe and to the United States. The railroads created a major change in the way people and goods were moved. Modern medical practices and increasing wealth caused a large increase in the world's population. The path toward widespread use of electricity began in the 19th century with the work of Thomas Edison and Nikola Tesla. In 1859, Charles Darwin, an English naturalist born in 1809, published *On the Origin of the Species*, in which he outlined the concept of species evolution through natural selection. Darwin used the theory of evolution to explain how life on Earth developed and evolved into what we see today through survival of the fittest by adaptation to the environment. Darwin, who had become an agnostic, was criticized by those who advocated the Christian explanation of species existence, and the debate on teaching evolution in schools persists today, particularly in the United States. The theory of evolution is one of the greatest achievements in modern biological science.

There were, of course, great advancements in mathematics as well. By the 19th century, the nature of the work of the mathematician had evolved to something that we would recognize today. No longer were public mathematical competitions of the last several centuries the major outlet of the mathematician, but rather mathematicians began teaching in universities while conducting research and writing papers. By the 19th century, mathematicians would specialize in areas such as analysis, algebra, or geometry, which was quite different from the general mathematicians of prior centuries. It was during this century that the ideas for abstract algebra developed. Until this time, algebra primarily was concerned with solving equations.

Another significant development was in statistics, which would build from the advances in probability theory over the last several centuries. It was during this time that possibly the greatest mathematician who ever lived, Carl Friedrich Gauss, did much of his work in mathematics.

Carl Friedrich Gauss, born in 1777 in central Germany into a poor family, is often considered the greatest mathematician who had ever lived. He rounds out our top three mathematicians, alongside Archimedes and Newton. His major contributions were in number theory, analysis, geometry, and statistics. He has been called the "Prince of Mathematics," and he considered mathematics to be the "Queen of the Sciences" with number theory as the "Queen of Mathematics."

There is a legend concerning Gauss, who was a child prodigy, that when he was 3 years old, he had mentally corrected an error his father made while calculating his finances. Another legend says that when Gauss was around 7 years old, his teacher gave him the assignment of finding the summation of the positive integers 1 to 100. Within a few minutes, Gauss presented his answer as 5050. Completely amazed, the teacher could not understand how he found the answer so quickly. Gauss noticed that by looking at the far ends of the summation, $1 + 2 + 3 + \cdots + 98 + 99 + 100$, he would have 50 pairs of 101. In other words, we find that $1 + 100 = 101$, $2 + 99 = 101$, $3 + 98 = 101$, until we reach $50 + 51 = 101$. This meant that all Gauss needed to do was multiply 50 by 101 to yield 5050. This summation is an arithmetic series, which is a series of numbers such that the difference of any two consecutive terms gives us a constant difference. We could represent this as follows: $a_1 + (a_1 + d) + (a_1 + 2d) + \cdots + (a_n - 2d) + (a_n - d) + a_n$, where $a_1$ is the first term and $a_n$ is the last term. In the example from Gauss, the first term is 1, the last term is 100, and difference is 1. We can call the sum $A$. We can reverse the series to have $a_n + (a_n - d) + (a_n - 2d) + \cdots + (a_1 + 2d) + (a_1 + d) + a_1$. If we add the two series together, we see the $d$'s cancel and we yield $2A = (a_1 + a_n) + (a_1 + a_n) + (a_1 + a_n) + \cdots + (a_1 + a_n) + (a_1 + a_n) + (a_1 + a_n) = n(a_1 + a_n)$ because there are $n$ groups of $(a_1 + a_n)$. We divide both sides by 2 to yield $A = n(a_1 + a_n)/2$. Using the example from Gauss, we see that the sum, which is $A$, is $100(1 + 100)/2 = 50(101) = 5050$, the same result Gauss had found as a child. Also, recall that Aryabhata in India during the 5th and 6th centuries also worked with the arithmetic series.

In 1795, Gauss began his studies at the University of Göttingen, but he did not finish his degree there and instead completed his doctorate at University of Helmstedt in 1799. At the University of Göttingen, Gauss met a fellow student named Farkas Wolfgang Boylai, born in 1775 in Transylvania, which was in Hungary at the time. Boylai would also become a famous mathematician and remain Gauss' lifelong friend. At the age of 21, Gauss wrote *Disquisitiones Arithmeticae*, or *Number Research*, which can be considered for number theory what Euclid's *Elements* was for geometry. *Number Research* was not published until 1801. This book compiles the work of previous mathematicians in number theory, such as Fermat and Euler, into a comprehensive work. It more fully developed the proof given by Euclid of the fundamental theorem of arithmetic, which states that any integer greater than 1 can be written as the unique product of prime numbers. As stated earlier in Chapter 12, Gauss further built on the work done by Fermat and Euler in modular arithmetic and gave us the familiar modular congruence symbol $\equiv$, due to the similarity between congruence and equals. Gauss also showed how to construct a heptadecagon, a 17-sided regular polygon, using a compass and straight edge. Gauss' organized manner of stating theorems, providing proofs for those theorems, and then following with numerical examples set the precedent for other mathematicians to follow.

Gauss provided a proof for the fundamental theorem of algebra in his doctoral dissertation in 1799, which states that every polynomial has as many complex roots as its degree. Earlier, Lagrange, Euler, and Laplace had attempted to find the proof as well. Gauss produced two more proofs for the theorem in 1816 and another version of the original in 1849.

Today, we refer to a method of solving systems of linear equations as Gaussian elimination. However, as mentioned in Chapter 7, the ancient Chinese knew of this method as found in the *Nine Chapters on the Mathematical Art*. Further, Newton had applied this method to his work in 1670, which later appeared throughout European publications. Gauss gave us the modern notation for systems of equations, which is why the method finally was named after Gauss. However, it was actually a mistake in the historical facts that led to the labeling of Gaussian elimination. For matrices, Gauss–Jordan elimination is named partially after Gauss due to its reliance on the Gaussian elimination method. A simple example of Gaussian elimination is presented for solving the following system of linear equations:

$$3x + 5y = 11$$
$$2x + 6y = 10$$

$$-2[3x + 5y = 11]$$
$$3[2x + 6y = 10]$$

$$-6x - 10y = -22$$
$$6x + 18y = 30$$

Adding the two equations together yields

$$8y = 8$$
$$y = 1$$

$$3x + 5(1) = 11,$$

which means that $x = 2$. The solution is (2, 1).

Gauss made great contributions to statistics. In 1801, he predicted when the dwarf planet Ceres would reappear on its way around the Sun, by using the method of least squares. He outlined his work on the method of least squares in 1809. To assist in this problem, Gauss developed the normal distribution curve, sometimes called the Gaussian curve. In French-speaking places, this is sometimes referred to as the Laplace curve since Laplace had also discovered the normal distribution curve in his work with the central limit theorem in 1810. From 1989 until 2001, before the replacement of German currency with the euro, Gauss was honored by appearing alongside his normal distribution curve, or Gaussian curve, on the German 10 Deutsche mark currency note.

In the early 19th century, Gauss' father died and this was followed shortly after by his wife's death during childbirth. Then Gauss' son died. These events led Gauss to experience a deep depression. Moreover, the string of painful events continued as his second wife also died well before he did in 1831. There is a legend about Gauss that said while working on a problem, he was interrupted to be told that his wife was dying and he replied, due to his obsession with his work, that she should wait until he was finished. Gauss died in 1855.

Before we leave Gauss, the topic of non-Euclidean geometry should be examined. Gauss' friend Farkas Wolfgang Boylai tried to prove Euclid's parallel postulate from Euclid's other axioms for many years, but he unfortunately did not succeed. In 1802, he had a son, Janos, who continued to work in this area. Janos Bolyai began his study of mathematics at an early age due to his father's influence, and like his father, he also developed a keen interest in Euclid's parallel postulate. His father warned him not to become obsessed with the parallel postulate and that, like the sensual passions of sexual intercourse, it would deprive him of his time, health, and happiness. Because the Bolyais were not wealthy, Farkas could not afford to send his son to receive advanced mathematics instruction, which meant that Janos studied engineering at the Royal Engineering College in Vienna, Austria, from 1818 to 1822, and then entered a military career with the army engineering corps. In the 1820s, Janos discovered non-Euclidean geometry and published his work as an appendix to his father's book in 1832. Janos poetically said to his father that out of nothing he had created a strange new world. The non-Euclidean geometry that followed from not assuming the parallel postulate was named absolute geometry. On seeing this work, Gauss commented that to praise the son's work would be to praise himself because Gauss claimed that he had also discovered these same results at the turn of the 19th century. It is said that Gauss refrained from publishing this work because he feared that it would not be accepted. This disdain toward Janos' work strained the friendship between Gauss and Farkas. However, in a commentary to a friend, Gauss said he considered the son's work to be quite brilliant.

In addition to mathematics, Janos is remembered for having a talent for languages: he learned nearly 10 different languages, including Chinese. He was also known for never smoking, drinking, or even taking coffee. Janos died in 1860. Although we have no actual pictures of Janos, the Hungarian postal service released a stamp in his honor in 1960 (the picture on the stamp is not actually Janos' picture). A similar interesting story exists for 18th- and 19th-century French mathematician Adrien-Marie Legendre, who was born in 1752 and died in 1833. Until 2009, for nearly 200 years, the mathematics community thought the picture of French politician Louis Legendre was a picture of Adrien-Marie Legendre because the caption to the picture listed the surname only.

Later, Gauss found that Nikolai Lobachevsky independently developed his own non-Euclidean geometry and published it in the same year as Janos, 1829. Again, Gauss stated that he worked on this for many years. It should be observed that Islamic mathematician Khayyam worked in non-Euclidean geometry in the 11th century, and Saccheri and Lambert worked in this area in the 18th century as well. However, Janos and Lobachevsky published works that much more fully expanded on absolute geometry by not assuming the parallel postulate.

Nikolai Lobachevsky, born in western Russia in 1792, attended Kazan University until 1811; in 1814 he was appointed as professor and in 1827 as an administrator. It was during his time studying at Kazan University that he was taught by a professor who had previously taught Gauss. He published his work in 1829 in his university's journal, *Kazan Messenger*. However, his work was rejected by the St. Petersburg Academy of the Sciences. In fact, it took decades for non-Euclidean geometry to be accepted as legitimate, partially due to the obscurity of the publications in which both works appeared. Lobachevsky died in 1856.

It has been debated if it was Gauss, and not Boylai and Lobachevsky, who had discovered the properties of non-Euclidean geometry in

19th-century Europe. Given that Gauss was close friends with Boylai's father and that Lobachevsky was taught by a former teacher of Gauss, it is very likely that both Boylai and Lobachevsky were influenced by Gauss' ideas. Boylai and Lobachevsky both published their work on non-Euclidean geometry, while Gauss did not. Hence, Boylai and Lobachevsky received the credit for discovering non-Euclidean geometry in 19th-century Europe. After over 2000 years, it was finally realized that Euclid was correct in regarding the parallel postulate as a postulate and not as a theorem because without assuming the parallel postulate, a different kind of geometry would have been developed. If it were truly a theorem, then the geometry of Boylai and Lobachevsky would have been consistent with Euclid's geometry.

The specific name of the geometry that Bolyai and Lobachevsky discovered is called hyperbolic geometry, which is the geometry inside a circle. In hyperbolic geometry, there are infinitely many lines through a point that are parallel to another line not on that point, unlike Euclidean geometry with the parallel postulate stating there is only one such line. In Euclidean geometry, all triangles have the sum of their interior angles equal to 180°, but in hyperbolic geometry, the sum of the interior angles is less than 180°.

The most famous mathematician who was a student of Gauss is probably Bernhard Riemann, who contributed to many areas of mathematics including non-Euclidean geometry, calculus, and analysis. He was born in Hanover, Germany, in 1826, and exhibited mathematical skills early in life. He began studying at the University of Göttingen in 1846, and initially studied theology, but later switched to mathematics. He left for Berlin University in 1847, but returned to the University of Göttingen in 1849 to complete his doctoral dissertation under Gauss.

In 1856, Riemann developed a new non-Euclidean geometry called elliptic geometry, which is the geometry on a sphere such as the Earth. In this geometry, the parallel postulate can be restated as follows: For a line and a point not on the line, there are no other lines that could contain the point that is parallel to the first line. In this geometry, a line would be a great circle, also known as a Riemannian

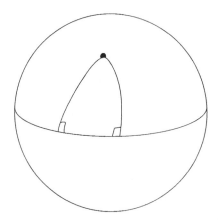

**FIGURE 14.1**   Elliptical Geometry Triangles

circle, which is a circle that has its diameter, the diameter of the sphere. The equator is an example of a great circle. A line segment would be an arc of a great circle, which means the shortest distance between any two points is an arc. In elliptical geometry, the sum of the interior angles of triangles exceeds 180°, and two lines, or great circles, intersect in two points. See Figure 14.1.

Interestingly, in Jules Verne's 1873 book, *Around the World in Eighty Days*, Verne's character, Phileas Fogg, takes a wager that he could circumnavigate the world in no more than 80 days. However, the entire trip takes place above the equator as Fogg travels from London through Europe to Egypt, India, Singapore, China, Japan, the United States, Ireland, and back to London. Since Fogg did not pass a great circle, he technically did not circumnavigate the world. If he would have gone a little farther south of Singapore he would have completed a great circle. If this is difficult to understand, imagine Fogg walking around the North Pole and claiming to have circumnavigated the world. Keep increasing the radius from the North Pole and it can be seen that not until the equator is reached would a true circumnavigation take place.

In 1859, Riemann devised a conjecture called the Riemann hypothesis, which states that the nontrivial zeros on the real part of a certain type of function called the Riemann zeta function have a value of 1/2.

The implication of the Riemann hypothesis involves the distribution of prime numbers and had a huge impact on number theory. Indeed, many other theorems are dependent on the Riemann hypothesis. Because of the proof of Fermat's last theorem presented in 1995, we can now consider the proof of the Riemann hypothesis to be the new "Holy Grail" in mathematics. In 2000, the Riemann hypothesis was one of seven Millennium Prize problems worth US$1,000,000 from the Clay Mathematics Institute, a nonprofit group in Cambridge, Massachusetts. It is still unproved today.

Calculus students will be familiar with Riemann's work when they encounter Riemann sums, which involve breaking the area under the curve into small rectangles to approximate the area under the curve. As the rectangles become smaller and smaller, we approach the true area of the definite integral. The Riemann integral is named after Riemann and he was innovative in its rigorous approach to the integral using a limit perspective.

Riemann traveled to Italy for his health for many years. Unfortunately, in 1866 he died there of tuberculosis, before his 40th birthday. His housekeeper found many of his unpublished papers. It is believed that Riemann preferred not to publish work that was unfinished.

Next, we shall look into the lives and accomplishments of three female mathematicians in the 19th century. The first mathematician was a frequent correspondent with Gauss named Sophie Germain, who was born in Paris in 1776. She became interested in mathematics at the age of 13 on reading about the death of Archimedes. Despite her parents' disapproval and disposition that a girl should not study mathematics, Germain read as many mathematics books as possible in her father's library late at night while wrapped in warm blankets (her parents took away her warm clothing, in an attempt to keep her in bed instead of reading mathematics). They eventually came to terms with her passion to study mathematics.

Unable to formally study mathematics as a woman, Germain studied the lecture notes of courses at École Polytechnique, including the work of Lagrange. She submitted her work to Lagrange under the pseudonym M. LeBlanc in an attempt to hide her identity as a woman. Lagrange found out the truth but still respected her work and became her mentor. Germain also wrote to Gauss using the name M. LeBlanc. In the early 19th century, the French army occupied Gauss' town. Germain feared that Gauss would suffer the same fate as Archimedes so she contacted the French commander, a friend of her family, to take special concern for Gauss. On learning this, Gauss also learned that M. LeBlanc was a woman, but like Lagrange he deeply respected her.

In 1815, Germain won a prize for her work in elasticity from the Paris Academy of Sciences in a mathematics competition. She did not show up to receive the prize because she felt that as a female mathematician, she would be unappreciated. However, after the competition, Germain came to realize that the area she preferred was number theory. The Paris Academy of Sciences offered a prize for finding a proof for Fermat's last theorem after the 1815 contest. This led to one of Germain's most important contributions through her work on Fermat's last theorem. She probably made more progress toward the proof than anyone else did in the 200 years of attempts. It is speculated that Germain could have been even more productive if she were permitted to formally study mathematics. Despite the restrictions placed on women, Germain's accomplishments are nonetheless very notable. Germain died of breast cancer in 1831, at 55 years old.

The next mathematician is Sofia Kovalevskaya, born in 1850 in Moscow, Russia. She developed an interest in mathematics at an early age and benefited from lessons provided by the family tutor. However, a short time later, her father decided he no longer wanted her to study mathematics, so like Germain, Kovalevskaya discreetly studied mathematics late into the night. Eventually, her father, like Germain's, allowed her to study mathematics when he realized that she was gifted in the subject. In 1868, Kovalevskaya married Vladimir Kovalevski, who later worked with Charles Darwin. Thus married, she would have a husband who could grant her permission to leave the country to study, which was required in Russia at the time. She was forced to go abroad since women were not allowed to study in universities in Russia.

In 1869, she attended the University of Heidelberg in Germany. However, as a woman, she was permitted only to audit the classes. Due to the presentation of her work on partial differential equations and her work in other areas, she became the first woman in Europe to receive a doctorate in 1874. It was from the University of Göttingen that she received her degree. Because of her difficulty finding a professorship, she eventually took a position at Stockholm University in Sweden in 1884. While at the University of Sweden, she did work in analysis and was the editor of a mathematical journal. She separated from her husband in 1881, and he committed suicide in 1883. She died of influenza at 41 years old in 1891.

The third prominent woman of the 19th century to work in mathematics was the writer Ada Lovelace. Lovelace, born in London in 1815, was the daughter of Lord Byron. She grew up without her father, and unlike the previous two women, she was encouraged to study mathematics by her mother. One of her tutors was mathematician Augustus De Morgan. In 1833, she met Charles Babbage, who is considered to be the "Father of the Computer" due to his designs of the Difference Engine. Babbage, born in 1791 in London, later designed his Analytic Engine, the successor to the Difference Engine. Although both computers were designed but not actually built by Babbage, Lovelace annotated Babbage's notes creating the world's first computer program. Lovelace can be considered the "Mother of Computer Programming." Lovelace died of cancer in 1852 at the young age of 36. Babbage is remembered as part of the Cambridge University group who advocated for the Continental calculus notation of Leibniz as opposed to the Newtonian notation, which was held sacred in England. Babbage died in 1871.

A special note should be made on female contributions to mathematics up until the 19th century. So far, we have addressed only five female mathematicians among the many male mathematicians. We began with the wife of Pythagoras, Theano, and then discussed Hypatia. We should be reminded that in terms of progressiveness in attitudes toward women, the Pythagoreans were very much ahead of their time. Throughout history, women faced extreme challenges in entering the field of mathematics or studying other subjects due to the restrictions universities placed on women. It is therefore unfair to judge female accomplishments against those of males in mathematics due to the great disadvantage in educational attainment that women faced even as late as the 20th century. It can be argued that even in our modern time, there persist cultural beliefs about women and mathematical ability that hinders female interest and entry into the formal study of advanced mathematics. Changing these attitudes is an important task that we have not yet fully accomplished, by any means.

One of the most prolific mathematicians of the 19th century was Augustus Louis Cauchy. Cauchy contributed to the development of analysis, and he introduced the idea of limits and continuity into calculus. He contributed to the perspective that the integral could be defined by the limit of regions under a curve and used the limit definition familiar today in elementary calculus for the derivative, along with using Lagrange's notation. Cauchy was born in Paris in 1789, and his family, who were supporters of the monarchy, fled Paris during the French Revolution to avoid the chaos. Cauchy became an engineer in 1810, but became a professor of mathematics in 1815 at the École Polytechnique in Paris. In 1830, King Charles X of France was facing abdication due to the sentiments in Parliament. His reaction was to tighten his grip on the government, which sparked the July Revolution and led to student rebellion. During this time, Cauchy fled France to Switzerland and lost his position at École Polytechnique. He later returned to university teaching in Paris, and died in 1857.

Next, we turn to the French mathematician Évariste Galois (pronounced "Gail-wah"), who lived an even shorter life than his Norwegian contemporary, Niels Abel. Abel, for whom the Abelian group in abstract algebra is named, died of tuberculosis at the age of 26 in 1829. Galois was born in 1811 outside Paris and died at the age of 20 in 1832. Like Abel, Galois worked in abstract algebra. The work of Galois and Abel demonstrates their brilliance, in which they had accomplished so much with such short lives.

In 1828, Galois was rejected by the École Polytechnique, but was accepted during the same year at the École Normale, which, at that time, was oriented toward teacher preparation and was not as prestigious as the École Polytechnique. Despite this, in the 20th century, the École Normale produced many recipients of the prestigious Fields Medal in mathematics. Galois did early work in abstract algebra and was early in using the term "group." He developed what is referred to today as Galois theory in abstract algebra.

The students at the École Normale were kept locked in the school during the July Revolution in 1830, and Galois sharply criticized this decision, which led to his expulsion and his entrance into the anti-monarchy National Guard militia. After the July Revolution, Louis-Philippe became king. At a celebration honoring members of the National Guard who had been arrested, Galois made a toast in the new king's honor while holding a dagger over his glass. For this threat against the king, he was arrested. On his release, he was arrested again for wearing a uniform of the National Guard. During his half-year sentence in prison, he continued his work in mathematics.

On May 30, 1832, Galois was involved in a duel that may have been over a woman he loved who was the daughter of the physician at the prison where Galois was serving his time. However, it is possible that this is a cover story for the government's action against him. Sensing impending doom, Galois allegedly stayed awake all night writing his final thoughts, including much of his work on group theory. On May 30, Galois was shot by his adversary and he died the next day. His dying words were to his brother, Alfred, whom he begged not to cry for him because Galois needed all of his courage to die at the young age of 20. One could only ponder how productive he may have been had he lived a longer life.

Let us move on to the development of statistics, which would see its early development during the 19th century. Earlier in this chapter, we discussed Gauss' use of the method of least squares, which would later be applied not only to natural phenomenon such as the movement of heavenly bodies but also to the social sciences. Recall that Gauss had worked with the normal curve, sometimes called the

Gaussian curve, which became very influential in later statistical work. Statistics developed on the foundation of probability theory, which had been advancing over the last several centuries.

Francis Galton, half-cousin of Charles Darwin, was born in 1822 in Birmingham, England, and died in 1911. He studied medicine and mathematics at Cambridge University, but after his father died, he had enough money to take his focus off his education and career and instead direct it toward his passion for travel. He is known to have traveled widely around Europe, Africa, and the Middle East. He wrote a guide for explorers called *The Art of Travel*. In addition to his global explorations, he worked in many different fields, including biology, psychology, and meteorology. However, for our purposes here, he is best known for his work in statistics.

Galton developed the use of survey research using questionnaires, and he applied his work in statistics to intelligence testing. He developed the concept of statistical correlation, which measures the relationship between two variables. For example, if we wanted to see the relationship between mathematics achievement and attitudes toward the subject, we could gather data to determine the quantitative relationship between the two variables. Research studies have found that there is a relationship between achievement and attitudes in mathematics. It is important not to necessarily interpret correlation with causation. One variable could cause the other, or they might both be caused by a third confounding variable. Perhaps higher achievement in mathematics leads to better attitudes, or perhaps better attitudes lead to higher achievement in mathematics. Perhaps there is a third variable we have not considered.

Karl Pearson is our next great contributor to 19th-century statistics and is considered to be one of the founders of modern statistics. Pearson, born in 1857 in London and died in 1936, had improved on the correlation work of Galton, and the Pearson coefficient that measures correlation is named after him. Pearson studied mathematics at Cambridge University and conducted graduate work in physics at the University of Heidelberg in Germany. He went on to hold positions at several universities throughout his career.

Pearson did work with the standard deviation, which measures the spread of data, and also coined the term "standard deviation." Gauss had earlier called this the mean error. For an example of the use of standard deviation, let us say that two classes both have average test scores of 80. We may at first conclude that the classes are identical. However, on closer examination, we may realize that they are quite different since in one class, most students may have scores centered around 80, while in the other class, half of the students may have scores around 100 and the other half have scores around 60. The first group would have a smaller standard deviation than the second group.

Pearson is credited with developing statistical hypothesis testing in inferential statistics, including $p$-values and the $\chi^2$ (chi-squared) statistical test. An inferential statistical test finds the probability of an outcome, and if the probability of the outcome occurring by chance alone is too small, we conclude that the outcome best describes the situation. For example, imagine we want to measure the test scores between two classes. If the class average for one group is 84 and the average for the other group is 76, we may want to know if this is a real statistical difference. If we calculate the probability that this difference occurred by chance alone, and we find that the probability is sufficiently small, 0.01, for example, we conclude that there is a real statistical difference between the two classes.

We shall end this chapter with an interesting story that connects mathematics with map making. In 1852, Francis Guthrie, who was the brother of Fredrick Guthrie, a student of Augustus De Morgan, was coloring the counties of England on a map and realized that only four colors were needed. This was communicated to De Morgan, who was a British mathematician born in British India in 1806 and would be later known for De Morgan's laws, related to the work of William of Ockham in the 12th century. The necessity of using at most four colors to color the regions of any map so that adjacent regions would have different colors would become known as the four-color theorem. Numerous attempts at proof were attempted, but the actual proof came in 1976 from Kenneth Appel and Wolfgang Haken at the University of Illinois, Urbana-Champaign, using a computer. This was the first major theorem to be proved using a computer, but not verifiable by mathematicians using traditional methods of proof. This was very controversial since some mathematicians do not accept this form of proof.

In the next part of the book, we shall explore the 20th and 21st centuries in mathematics and especially the shift from Europe to North America because of the events happening in world history. During World War II, many mathematicians and scientists escaped Europe for North America to flee from Nazi Germany. The divergence in prosperity between Europe and the United States after the war attracted more mathematicians and scientists, which again added value to the United States by bringing many of the best in the world to the shores of America. Much of Europe experienced a "brain drain" during this period. We shall explore this history as well as the history of mathematics education in the United States in the next and final part of this book.

# MATHEMATICS IN THE 20TH AND 21ST CENTURIES

PART 4

# MATHEMATICS IN THE 20TH AND 21ST CENTURIES

# CHAPTER 15

# EUROPEAN, AMERICAN, AND GLOBAL MATHEMATICS: THE 20TH AND 21ST CENTURIES

The 20th century saw many great changes. The new unified state of Germany grew dramatically in power in the beginning of the century and challenged the long held hegemony of the United Kingdom and France. Furthermore, World Wars I and II brought the world two devastating wars, the second more deadly than the first, in the first half of the 20th century. The age of great empires and colonialism came to end as the Ottoman and Austro-Hungarian Empires collapsed, and by the end of World War II, an exhausted British Empire gave up most of its overseas territories, as did the other great powers in Europe, such as France. In between the two world wars, the world entered the global financial crisis of the Great Depression. The threat of a third, nuclear world war was manifested in the Cold War between the world's remaining superpowers, the United States and Soviet Union, and remained so until the collapse of the Soviet Union. The Soviet Union was the product of the Russian Revolution at the end of World War I and existed as a state throughout much of the 20th century, until it collapsed in the final decade of the century. Instead

*The Development of Mathematics throughout the Centuries: A Brief History in a Cultural Context*, First Edition. Brian R. Evans.
© 2014 John Wiley & Sons, Inc. Published 2014 by John Wiley & Sons, Inc.

of World War III, many bloody proxy wars were fought around the world between the United States and Soviet Union. It is during the 20th century that war had become mechanized through advanced technology. Although technology benefited the world greatly, issues of environmental degradation resulting from the widespread use of technology grew to be a serious issue in the late 20th century and remain so today.

While the 20th century experienced large-scale hot and cold wars that involved major powers such as the United States, Soviet Union/ Russia, United Kingdom, France, Germany, Italy, and Japan, we could characterize the early 21st century as a period of conflicts involving the Western and Islamic worlds and primarily centered around a small but influential group of fundamentalists. While traditional wars had been a constant theme throughout the 20th century, the early 21st-century conflicts involved nonstate belligerents utilizing terrorism. The global dependence on the limited and diminishing resource of oil, coupled with large-scale population growth and development, has led to a shift in the power dynamic because of our growing demand. Since the late 20th century, the oil-rich regions of the world, such as the Islamic Middle East, are recognized as very important areas. Furthermore, the Arab Spring, a popular uprising in many Arab countries, has changed the political landscape of the Middle East. An additional strain for the world has been the global financial crisis that emerged in the first decade of the 21st century, considered to be the worst financial crisis since the Great Depression.

Despite the major calamities of the 20th century, scientific achievements and progress have dramatically changed life over the course of this century in a way to which the experiences of earlier centuries cannot possibly be compared. In 1903, the Wright brothers in the United States built the world's first successful airplane. By World War II, airplanes were an integral part of the war, with air power replacing the role that sea power had previously maintained. By the end of the war, Germany built the world's first jet engine, and by 1957, the Soviet Union sent the world's first artificial satellite into space. *Sputnik 1* began the Space Race between the Cold War rivals, as will be addressed in the next chapter. The Soviet Union sent the first human, Yuri Gagarin, into orbit in 1961. The United States followed by sending the first humans to the Moon in 1969; Neil Armstrong became the first human to walk on the Moon. The advent of aviation allowed people to move around and to travel the world with amazing speed at low cost.

The 20th century experienced much advancement in science and technology. In the 1920s, a Belgium priest and physics professor, Georges Lemaitre, developed the "big bang" theory that explained the formation of the universe. This discovery eventually did for physics what Darwin did for biology in the previous century. There was no longer a need to rely on religion to explain the existence of life or the universe. World War II led to fierce competition and the development of many new technologies, including the first computers and nuclear energy and bombs. During the mid-20th century, food production became vastly more efficient than it previously had been because of the application of science and technology in agricultural techniques. This Green Revolution led to an astronomical increase in the global population, along with incredible advances in medical science including the widespread use of antibiotics and vaccines. Although there were great hardships during the wars of the 20th century and poverty persisted worldwide, the standards of living greatly increased in Western Europe and North America during the 20th century. The United States experienced more prosperity in the second half of the 20th century than had ever been encountered in history before. Daily living in wealthy nations changed dramatically over the course of the 20th century with the expansion of the middle class. The 20th century experienced widespread access to the mass production of many products, including the automobile, beginning with Henry Ford's Model T in the early 20th century, and other household appliances. Moreover, the use of fossil fuels and electricity directly impacted the increased standard of living.

The advancements in aviation and widespread automobile ownership increased dramatically in the first half of the 20th century, but slowed down during the second half of the century. Nevertheless, near the end of the 20th century, new technologies altered the world in ways that were not imaginable before. The personal computer and the Internet allowed people to communicate and disseminate knowledge at incredible speeds, and created a more level playing field in global competition as humanity entered the 21st century. As earlier addressed in this book, the knowledge dissemination produced by the Internet will likely rival the revolution in knowledge dissemination brought about by Gutenberg's printing press in the mid-15th century, which led to a more efficient dissemination of knowledge through printed books. We have yet to experience the full potential of what access to this massive amount of knowledge and efficiency can do for the world. Globalization in the late 20th and early 21st centuries has led to great advancements and has propelled many nations forward such as China, India, Brazil, and Russia.

Great scientific and mathematical advancement came about during the 20th century. The career path of mathematicians and professors became much more common in the early 20th century with more graduate students receiving doctoral degrees in mathematics. The tone for 20th-century mathematics was set in 1900 at the second International Congress of Mathematicians in Paris, the largest mathematics conference in the world that first began in Switzerland in 1897 and currently is hosted by the International Mathematical Union. David Hilbert presented 23 problems to the international mathematics community at the conference. These problems became the famous Hilbert problems that are still being worked on today. Over the 20th century, a number of these problems were solved, but many are yet to be solved. One of the more famous problems is the Riemann hypothesis, Hilbert's eighth problem, described in the previous chapter. Another problem addressed an issue in set theory proposed by Georg Cantor, a German mathematician born in St. Petersburg in 1845 and died in 1918. Cantor is considered to be the "Father of Set Theory," and worked with an idea on infinity that would lead Paul Cohen, born in 1934 in New Jersey in the United States and died in 2007, to show that there can be two types of mathematics based on our understanding of infinity and Cantor's continuum hypothesis, which also happens to be Hilbert's first problem.

Hilbert, born in 1864 in Königsberg, Germany, the location of Euler's Bridges of Königsberg problem, graduated from the University of Königsberg in 1885 and went on to work at the University of Göttingen. He contributed to many areas of mathematics, including abstract algebra and geometry. He also developed a new consistent axiomatic system for Euclidean geometry. During the rise of Nazism in the 1930s in Germany, many Jewish faculty were removed from the mathematics department at the University of Göttingen. When seated next to the Minister of Education at a dinner, Hilbert was asked by the minister how mathematics at the University of Göttingen was doing after removing Jewish influence. Hilbert replied that there really was no mathematics at the university any longer. Hilbert died in 1943, and the words on his grave echo his retirement speech in 1930, which said there are no unsolvable problems. He famously said that we must know and that we will know. However, this idea of being able to solve every mathematical program experienced a derailment with Kurt Gödel's incompleteness theorem, which is really two theorems in mathematical logic that state there are limitations in almost all axiomatic systems. That is, there are propositions that cannot be proved or disproved in the axiomatic system being used.

Gödel was born in 1906 in Brunn, which today is in the Czech Republic. In 1940, he fled to the United States from Nazi-controlled Europe and took a position at Princeton University. At Princeton's Institute for Advanced Study, he became good friends with Albert Einstein. Einstein, born in 1879 in Germany, also escaped the Nazis in 1933 and took a position at Princeton University. Einstein, also a Pythagorean in terms of his vegetarian diet, published the *Theory of Relativity* in 1905, which relied on the mathematical developments in non-Euclidean geometry and revolutionized Newtonian physics that had been unchallenged for over 200 years. Einstein famously urged US President Roosevelt at the beginning of World War II to fund the development of atomic weapons before Germany did. In 1945, the United States dropped the world's first atomic bomb on Japan. During testing, Robert Oppenheimer, the "Father of the Atomic Bomb," famously recalled the words of Shiva in the *Bhagavad Gita* in which

Shiva declares that he has become death, the destroyer of worlds. Gödel died in 1978 and Einstein died in 1955.

The study of statistics, which had been developing in the late 19th century, continued to advance throughout the 20th century. Today, statistics is often used in many areas such as social sciences, medical sciences, and business, among others. The most prominent figure in 20th-century statistics development is Ronald Fisher, who was born in 1890 in London. Performing well in mathematics since he was a child, Fisher graduated from Cambridge University in 1912 with a concentration in mathematics, but with an interest in biology and evolution. After graduation, Fisher worked as a statistician and taught mathematics and physics. In 1917, a feud developed between Fisher and Pearson due to Pearson's criticism of Fisher's work. In 1918, Fisher published his work that contained the introduction to the analysis of variance (ANOVA). Although Pearson did not reject the work, he expressed some reservations about it. In 1922, Pearson criticized Fisher's use of the $\chi^2$-test. Ironically, when Pearson retired from his position at University College in London in 1933, his position was split into two and was subsequently held by Fisher and Pearson's sons. Fisher took a position at Cambridge University in 1943 and, in 1957, left for Australia, where he died in 1962.

A statistician who had a good relationship with Pearson was William Sealy Gosset, also known as "Student." Gosset was born in Canterbury, England, in 1876. He graduated from Oxford University in 1899 and took a position as a statistician at the Guinness brewery in Dublin. Because Guinness did not want its employees to be published, which could hurt the company, Gosset published under a pseudonym "Student." Today, we refer to the "Student's $t$-test" in statistics after Gosset's discovery. Gosset was interested in statistical testing of small samples to help him with his work. The Student's $t$-test allows us to determine if a real difference exists for a small sample, such as the kind Gosset dealt with at his job. Interestingly, while Gosset studied with Pearson in 1906–1907, he also had a good relationship with Fisher. He died in 1937.

Statistics became very important and useful during World War II, since determining the quality of equipment and soldiers as well as being able to make decisions regarding predictions in bombing accuracy and weather were important variables for military leaders. The 20th century made progress not only in statistical theory but also in the way statisticians and other users of statistics performed their computations. The electronic calculator made the statistical calculators easier, but computers later allowed statistics to be done using statistical software in the 1960s such as the Statistical Package for the Social Sciences (SPSS) and Statistical Analysis System (SAS), among other software packages. Even Excel, part of Microsoft Office, can perform fairly sophisticated statistical tests.

An important figure in mathematics in the 20th century was Bertrand Russell, who was born in 1872 in Southeast Wales. His grandfather was John Russell, who served as Prime Minister under Queen Victoria. Bertrand Russell studied at Cambridge University and taught there until 1916 when he was convicted and subsequently fired from his position for engaging in antiwar activities. Russell spent most of his life as an antiwar activist, and in 1955, he worked with Einstein to call for the elimination of nuclear weapons. In addition to his antiwar stance, Russell is remembered for his atheism and his 1927 work, *Why I Am Not a Christian*. Before World War II, Russell lectured in the United States and was given a position at the City College of New York. However, he lost his position due to public protest against Russell's liberal view of marriage and sexual intercourse. John Dewey, the prominent educator at the University of Chicago, was among those who came to Russell's defense. In 1944, Russell returned to Cambridge University. Despite history's remembrance of Russell for logic and mathematics, he won the Nobel Prize in literature in 1950. Russell died in 1970.

Russell is best remembered in mathematics for his work in logic and set theory. He published his *Principia Mathematica* in the early 20th century with A. N. Whitehead in order to develop a firm foundation for mathematics in set theory and logic. He is known for Russell's paradox, discovered by Russell in 1901, and also discovered by German mathematician Ernst Zermelo shortly prior. The paradox is based on the set theory of Cantor and states that we can specify a set $A$ that is the set of all sets that are not members of themselves. If $A$ is a member of itself, then we have a contradiction because $A$ is the set of all sets that are *not* members of themselves. However, if $A$ is not a member of itself, then $A$ would have to be member of itself since it would be a set that is not a member of itself, which $A$ would contain, which is another contradiction. Russell helped us understand this by giving more real-world examples of the paradox. For example, although this procedure is no longer the practice, barbers used to frequently shave men's beards. Imagine there is a town in which every man either shaves his own face or goes to the town's only barber. As a rule, the barber shaves every man, and only those men, in town who do not shave their own faces. If the barber shaves his own face, then we have a contradiction because the barber shaves only the men who do not shave themselves. Hence, he is not allowed to shave himself. If the barber does not shave himself, then the barber would have to shave himself, as the town's barber, since the barber shaves all men who do not shave themselves. This is another contradiction.

In 1935, a new mathematician was born in France who was unlike any other mathematician we have discussed so far. His name was Nicolas Bourbaki, and in the year of his birth, he began working toward writing a comprehensive book on mathematics founded on set theory in line with Russell's idea. The book he produced was called the *Elements of Mathematics* and it eventually was published in many volumes. The original goal was to write a book on analysis, but later, it was expanded with the intention to place all of mathematics on a solid foundation. How could such a talented individual begin producing in his first year of birth? Well, he was not one individual but rather the publishing pseudonym of a group of mostly French mathematicians at École Normale formed by French mathematicians Henri Cartan and Andre Weil. Later, the group consisted of nearly 20 mathematicians including Alexander Grothendieck, born in 1928 in

Berlin, raised in France, and famous for his contribution to algebraic geometry before giving up mathematics for his antinuclear weapons activism. The founders of the group, Cartan and Weil, were both born in France in the early 20th century, and both lived very long lives, 108 and 92, respectively. Like Russell, Weil did not believe in war, and Weil was subsequently arrested for not serving in the military. In 1942, Weil left Europe for the United States, where he spent most of the rest of his life.

The name "Nicolas Bourbaki" was chosen as a joke. Prior to forming this group, a student had played a prank to make other students believe he was a famous mathematician. He presented fake theorems named after French generals. The final one was named after General Charles Bourbaki, hence the inspiration for the surname. World War II interrupted this work, but after the war, Bourbaki continued to be productive for many years.

One of the more interesting mathematicians in the 20th century was Paul Erdös, who was born in Hungary in 1913. Erdös is best remembered for his eccentric ways and his record number of publications that surpassed even those of Euler. Erdös had over 1500 publications, which makes Erdös one of the most prolific mathematicians who ever lived. He is known for his work in analysis, number theory, logic, and probability. Erdös had a difficult childhood since he was born around the outbreak of World War I. His parents were mathematics teachers, but there were restrictions on Jews entering the university. Even still, Erdös earned a spot due to his high scores on an examination. He received his doctorate in 1934 from University Pazmany Peter. Due to rising anti-Semitism, Erdös did his postdoctoral work in England. Shortly before the outbreak of World War II, Erdös left Europe for the United States and worked at several US universities. In the early 1950s, Erdös was not permitted to reenter the United States, possibly due to the anticommunist frenzy of McCarthyism sweeping the United States at the time. He spent time in Israel until the US government allowed him to return during the early 1960s.

Erdös collaborated in the writing of his papers with many mathematicians. He was known to eschew material possessions and carried a single suitcase with him as an itinerant scholar who stayed in house after house of the various mathematicians of his time. He would show up and declared that his "brain was open," which meant that he was ready to coauthor a paper while staying at the house, and that only when he was finished would he move onward. Erdös often won mathematical prizes, but not needing much money himself, he donated a portion of it to help out other mathematicians. His primary interest was solving problems rather than proving theorems.

An interesting phenomenon arose regarding the Erdös number. Similar to the Degrees of Kevin Bacon game in which players try to determine how many "degrees" away an actor is from actor Kevin Bacon, mathematicians can find an Erdös number indicating how far away a person is from Erdös, given the large number of publishing coauthors Erdös had. Erdös is assigned an Erdös number of zero. Those he published with, over 500 people, have Erdös numbers of 1. Those who collaborated with the number 1 people, but not with Erdös himself, are given a number of 2, and so on. Interestingly, mathematician–actress Danica McKellar, who played Winnie Cooper on the television show *The Wonder Years*, has since written mathematics books to get adolescent girls interested in mathematics. She has Erdös and Bacon numbers 4 and 2, respectively. Since baseball legend Hank Aaron and Erdös both signed a baseball together at an honorary degree ceremony, some say that Aaron has an Erdös number of 1. Erdös died in 1996, and he was one of the greatest mathematicians of the 20th century.

In 1936, the first Fields Medal in mathematics was awarded at the International Congress of Mathematicians in Norway. It was founded by Canadian mathematician John Fields, and today awards Canadian $15,000 to the winner. The award is only available to mathematicians under 40 years old and is awarded every 4 years at the opening of the International Congress of Mathematicians. Many have considered it to be the "mathematics Nobel Prize." Alfred Nobel did not include mathematics as part of his Nobel Prize, and history does not know why. Some have speculated that a mathematician took the woman he loved, but the probable reason is that Nobel was not as interested in offering a prize in mathematics as he was in the other areas. The prize that is probably more analogous today to the Nobel Prize in mathematics is the Abel Prize, founded in 2002 by the Norwegian government and named after Norwegian mathematician Niels Abel. There was an initial interest in starting this prize in 1902, 100 years after Abel was born, but the prize was not founded until the bicentennial of his birth. This prize is awarded every year, and carries a much larger prize than the Fields Medal with the value at nearly US$1,000,000.

World War II led to many innovations to assist with the war effort. It is possible that the innovative work of cryptologists, mathematicians who work in the breaking of codes, greatly hastened the Allied victory in World War II. It is uncertain if the Allies would have lost the war without breaking the German code, but it is certainly true that the war would have been extended, possibly by several years, without the Allies breaking German code when they did. British Prime Minister Winston Churchill said that it was thanks to this work that the Allies won the war. The Germans had created an Enigma machine at the end of World War I that they used in World War II to communicate through encrypted code. Poland broke this code in 1932 and gave the intelligence, called Ultra, to the British and French. Ultra allowed the Allies to stay one step ahead of the German military maneuvers.

A notable cryptologist during the war was Alan Turing, born in 1912 in London. From an early age, Turing demonstrated great mathematical ability, but was often distracted by his own mathematical interests rather than that of his teachers. In 1930, his closest friend, Christopher Morcom, died of bovine tuberculosis, which led Turing to become an atheist. It is speculated that Turing had a romantic interest in Morcom. Turing's homosexuality would later create problems for him with the British government. He graduated from Cambridge University in 1934 and completed his doctorate at Princeton University in 1938. It was at Princeton that Turing designed the Turing machine for his doctoral dissertation. The Turing machine was not built, but rather served as a theoretical description of computations through the use of a strip of tape that would later influence the development of computer science. If a modern computer is Turing-complete, it means that the computer can simulate the single-tape Turing machine. After finishing his degree in the United States, Turing returned to Cambridge University and joined the cryptologists at Bletchley Park, the United Kingdom's primary cryptology station during the war.

In 1939, during his time at Bletchley Park, Turing developed the bombe, which was a device used to decrypt German code. Turing was in the United States between 1942 and 1943, which was around the time that the Germans increased the sophistication in their codes. During that time in the United Kingdom, the Colossus machine was designed in 1943 at Bletchley Park as a more sophisticated device to break the more advanced codes used by the Germans. The Colossus machine, designed by Tommy Flowers, utilized vacuum tubes and could be considered to be an early programmable electronic computer.

In 1941 in Germany, one of the world's first computers was built, called the Zuse Z3, designed by Konrad Zuse. In 1943, in the United States, another computer was built at the University of Pennsylvania commissioned by the US Army. The Electronic Numerical Integrator and Computer (ENIAC) was completed in 1946. Shortly after, the University of Pennsylvania built the Electronic Discrete Variable Automatic Computer (EDVAC), which used a binary system of numbers instead of decimal numbers. Today, all computers use binary operations. Toward the end of World War II, Turing was asked by the National Physical Laboratory in London to design a computer. In 1946, Turing completed his design for the Automatic Computing Engine (ACE), and the ACE was completed by 1950.

In 1952, Turing was accused by the British government of "gross indecency" for homosexual acts, which were illegal at that time in the United Kingdom. Turing was convicted and required either to go to prison or to undertake hormonal treatment to reduce his sex drive. Turing opted for the latter. However, in 1954, Turing died from cyanide poisoning. Many believed it to be a suicide, but his mother insisted that it was an accident. In 2009, British Prime Minister Gordon Brown finally apologized on behalf of the British government for its treatment of Turing, a person who should have been upheld as a national hero for his work. Today, Turing, along with Babbage, is considered to be the "Father of Computer Science."

John von Neumann, a Hungarian mathematician born in 1903, was an early pioneer in quantum mechanics, game theory, and computer science, among other areas. Like Erdös, he faced a challenge in attending college due to his Jewish background, but he showed great ability in mathematics and was granted a spot at the university. Unlike Erdös, he chose not to go to college in Hungary, but instead studied chemistry in Berlin and then Zurich. Von Neumann returned to Hungary and received his doctorate in 1926 in Budapest. He took on academic positions in Europe until he left for Princeton University in the United States in the early 1930s, where he joined the faculty of the Institute for Advanced Study with Einstein and Gödel. By the late 1930s, von Neumann began work on the Manhattan Project, and he was instrumental in the development of the atomic bomb. His work in this area led to further innovation in the early computers that were being developed. Von Neumann improved the way the computer worked in the memory processing, which influenced how computers are made today. In 1957, von Neumann died of cancer.

In the 1940s and 1950s, computer technology quickly advanced and began to replace older methods of mathematical computation. However, these computers were very large and expensive, so they were not practical for most people to own, including many scientists. In 1957, the same year as the Soviet Union's launch of *Sputnik*, the world's first digital compact calculator was introduced in Japan. However, this calculator was not portable. The first portable desktop calculator was introduced in the United Kingdom in 1961, followed by calculators in the United States. In 1967, Texas Instruments developed the first handheld calculator. By the 1970s, small, less expensive pocket calculators were being developed and mass-produced. By the mid-1970s, calculators became affordable for much of the population in the wealthy countries. In its earliest stages, calculators performed only the four basic functions of addition, subtraction, multiplication, and division. Later, other operations, such as the square root, were added. By the 1970s, scientific calculators appeared on the market that included many advanced functions that could handle operations such as trigonometric, logarithmic, and statistical calculations.

By the 1980s, the first programmable graphing calculators appeared, starting with those made by Casio and followed by those made by Hewlett Packard. In 1990, Texas Instruments began selling the TI-81. By the 1990s, computer algebra systems (CASs) were available in graphing calculators that allowed people to manipulate algebraic expressions such as factoring polynomials and performing matrix operations. CAS systems allowed one to find derivatives and integrals effortlessly. CAS had existed for computer applications since 1960s but developed into computer software such as MATLAB in the 1970s and Maple and Mathematica in the 1980s. Statistical software was developed in the 1960s with software packages such as SPSS and SAS. In the 1980s, interactive geometry software developed, such as the Geometer's Sketchpad. In the 21st century, school children now use "virtual manipulatives" on websites that are virtual versions of hands-on objects used for learning mathematics such as blocks, dice, and fraction tiles. Today, there is much debate regarding the benefits of these technologies for students. One argument is that the technology enhances education and frees students to explore deeper mathematics and problem solving. A counterargument is that students become too dependent on technology and do not get enough practice working out mathematics using paper and pencil.

In the previous chapter, the four-color theorem was introduced. Since the middle of the 19th century, many mathematicians attempted to provide a proof, but none succeeded. In the 1960s, mathematician Kenneth Appel, born in 1928 in Brooklyn, New York, suggested that a computer could potentially check the needed number of cases to verify the four-color theorem, still a conjecture at that point. Since Appel was not a computer programmer and the computers in the late 1960s were not powerful enough to complete this task, Appel announced that he would stop working on the problem in 1972. Programming expert Wolfgang Haken, born in 1932 in Berlin, approached

Appel and suggested working together so they might be able to prove the theorem using a computer, since there had been substantial progress in computing over the last several years. By 1976, Appel and Haken completed their proof using the computer. This was the first major theorem proved in this manner, although many in the mathematics community did not accept this form of proof. However, a change had occurred that was unprecedented throughout history. This was the first time in history that some mathematical proofs could be done by computer, but not verifiable in the traditional methods of proof that had existed for 2500 years. A simplified version was completed in 2005 that relies less on the confidence in the computer program, which increases our confidence in the validity of this process. Some may never accept computer-generated proofs and will accept only traditional methods, but more mathematicians have accepted computers as part of the work they do in both research and teaching.

We now turn to a darker side of recent mathematics history. Ted Kaczynski, also known as the Unabomber, was a mathematician who began killing and injuring people due to his antitechnology beliefs. Kaczynski was born in 1942 in Chicago, and his high intelligence and abilities in mathematics isolated him from other children. He finished high school early and entered Harvard University. While at Harvard, he participated in a psychological experiment that may have had long-lasting negative effects on him. Kaczynski graduated from Harvard in 1962 and became a graduate student in mathematics at the University of Michigan. In 1967, he joined the faculty at the University of California, Berkeley, but he left after 2 years because he did not perform well there, which was probably because of the requirement to teach undergraduate students. In 1973, he moved to rural Montana to live a simple life of self-sufficiency. His discontent with our technologically advanced society was reinforced as he witnessed the wilderness being commercially developed.

In 1978, Kaczynski sent his first homemade bomb to an engineering professor at Northwestern University. However, the suspicious professor reported the package to the police. Unfortunately, the explosion injured a security guard. Kaczynski followed with attacks on universities and the airline industry over the next 17 years while avoiding capture by federal authorities. Kaczynski extended his attacks throughout the years on universities and the airline industry to include the computer and timber industries. During this period, three people were killed and many more were injured.

In 1995, Kaczynski sent his manifesto, *Industrial Society and Its Future*, to the media demanding its publication in order for him to stop his attacks. The US government believed that publishing this work would lead to Kaczynski's capture, so it was published in the *New York Times* and *Washington Post*. While Kaczynski's brother had suspicions that his brother Ted was the Unabomber, the publication of the manifesto confirmed this, and Kaczynski's brother subsequently turned him in to the police. In 1996, Kaczynski was arrested and he was convicted to life in prison in 1998.

After exploring the dark side of mathematics, let us now move forward to discuss perhaps the greatest success in modern mathematics history. In Chapter 12, Fermat's last theorem was presented along with reference to the person who proved the theorem in 1995, Andrew Wiles. Wiles was born in 1953 in Cambridge, England, where his father was a professor of divinity at Cambridge University. At 10 years old, Wiles discovered Fermat's last theorem in a mathematics book and was intrigued that this was one of the few problems that could be understood by people without sophisticated knowledge of mathematics. He attempted to provide the proof but soon realized that he did not know enough mathematics for this task. Wiles finished his undergraduate degree at Oxford University in 1974 and his doctorate at Cambridge University in 1980. After graduation, Wiles took a position at Princeton University.

In 1986, Wiles heard about Ken Ribet, an American mathematician born in 1947 who was a professor at the University of California, Berkeley. Ribet proved a conjecture, now called Ribet's theorem, that says if a part of the Taniyama–Shimura conjecture were true, then Fermat's last theorem would follow. The Taniyama–Shimura conjecture was put forth by two Japanese mathematicians, Yutaka

Taniyama and Goro Shimura, and refined by Andre Weil. Interestingly, Taniyama committed suicide in 1958 under strange circumstances in which his suicide letter told us that the decision to kill himself had come to him suddenly. Wiles realized that if he focused his effort on proving the part of the Taniyama–Shimura conjecture that was needed, then he would complete a thread that would establish Fermat's last theorem. For 7 years, Wiles worked on this problem secretly in his attic, and then presented his work in 1993 at Cambridge University. He did not explicitly state the goal of his presentations, but at the end of his lectures, he wrote Fermat's last theorem on the board and quietly stated that he thought he would stop here. Although a fundamental unfortunate flaw in the proof was found, Wiles found a way around the problem with the help of his former graduate student in 1994, after almost giving up. In 1995, Wiles published his paper, "Modular Elliptic Curves and Fermat's Last Theorem" in the *Annals of Mathematics*. It is more than 150 pages long. Wiles received many awards for his work, and in 2000, he was knighted by Queen Elizabeth II. After over 350 years, and built on the contributions of many mathematicians, Fermat's last theorem was established.

In 2000, an American mathematician Stephen Smale, born in 1930, proposed 18 new problems for the 21st century in spirit of Hilbert's problems 100 years ago. However, these problems were not accepted with the same enthusiasm and fanfare with which the mathematics community accepted Hilbert's challenge. The first problem was the Riemann hypothesis, which was also included on Hilbert's list. The second problem was the Poincaré conjecture, proposed in 1904 by 19th- and 20th-century French mathematician Henri Poincaré. Both problems also are two of the seven Clay Mathematics Institute's Millennium Prize problems announced in 1999 in which the winner receives US$1,000,000. In 2002, reclusive and eccentric mathematician Grigori Perelman proved the Poincaré conjecture. Perelman was born in 1966 in St. Petersburg, Russia, at that time called Leningrad. His mathematics ability was apparent at a young age. He received the equivalent of a doctorate from the Leningrad State University.

Afterward, he worked at the Russian Academy of Sciences, at that time called the USSR Academy of Sciences. Due to his 2002 proof of the Poincaré conjecture, Perelman was awarded the Fields Medal in 2006, but he declined the award. Perelman thought that successfully finishing the proof was reward enough and he did not want recognition or monetary reward. In 2010, Perelman was awarded the Clay Mathematics Institute's Millennium Prize, which was worth US$1,000,000. However, again he declined. He thought that fellow mathematician Richard Hamilton should have received the prize with him. After proving the Poincaré conjecture, Perelman has since left mathematics due to his perceived disappointment with the mathematics community. He became unemployed and currently lives with his mother in St. Petersburg. When a journalist tried to reach Perelman, he responded by declaring that the journalist was disturbing him while he was picking mushrooms.

There has been more emphasis recently in the history of mathematics on the contributions of mathematicians who were female and/or non-white. Historically, it has been difficult for female mathematicians to develop their mathematics due to the overwhelming denial of educational opportunities for women needed to contribute substantially to mathematics. It was not until 1983 that the American Mathematical Society (AMS) had its first female president, Julia Robinson, born in 1919 in Missouri and died in 1985. Robinson also worked on Hilbert's tenth problem that was later solved in the Soviet Union in 1970.

In the history of the United States, the same denial for female access to mathematics had occurred for non-white people as well. In 1986, the African Mathematical Union Commission on the History of Mathematics in Africa (AMUCHMA) formed to promote the dissemination of knowledge of African mathematical accomplishments throughout history. The organization serves as a center for research in African mathematics history.

The Benjamin Banneker Association is an organization interested in promoting the mathematics education of African American students. Benjamin Banneker was a free African American mathematician born in 1731 in Baltimore. He was educated in a Quaker school when he was young, but taught himself mathematics later in life. In 1791, he assisted in the survey for the nation's new capital in Washington, DC. His writings consisted of the almanacs he had produced. Although a fire destroyed much of his work, there was one surviving journal that contained his mathematics problems. He died in 1806.

It was in the 20th century that African Americans began to have more access to formal mathematics education. Elbert Cox, born in 1895 in Indiana and died in 1969, was the first African American to receive a doctorate in mathematics. Euphemia Lofton Haynes, born in 1890 in Washington, DC, and died in 1980, was the first female African American to receive a doctorate in mathematics. The first African American inducted into the National Academy of Sciences was David Blackwell, who was born in Illinois in 1919 and died in 2010. His work led to the Rao–Blackwell theorem in statistics.

As emphasized several times throughout this book, we are experiencing one of the most significant periods in the history of human knowledge, not just for mathematics, but for all areas of knowledge. Just as the printing press allowed for the quick and efficient distribution of knowledge in the mid- to late 15th century, computers and the Internet are revolutionizing how people work, communicate, and share knowledge. Many people today cannot remember how they functioned before computers and the Internet. The author wrote this book with much greater ease than would have been possible in the past by using a laptop and having access to numerous online sources of information such as the MacTutor Mathematics History website at the University of St. Andrews. While relatively inexpensive air travel allows one to learn more by traveling the world, as the author has done, collaboration with others overseas is even easier now with online communications. SPSS has allowed the author to analyze data in his statistical work with unimaginable ease, and computer software allows us to visualize mathematics more easily than had been possible for several millennia. For those who use the Internet, which now make up almost a third of the world's population and are quickly increasing, access to knowledge for children and adults is at their fingertips. The ubiquitous use of electronic mail, social media sites, and smart phones has changed communication. Social networking has changed the way people interact and stay connected with one another. Online music, movie, media, banking, and travel booking, among other areas, have all transformed industries. The jobs we have and the way we work have evolved and continue to do so due to advancements in computer and online technology. Access to computers and the Internet has contributed to a major shift in global economics. It is now possible to send work from New York to Bangalore, India, through the Internet, where the work can be done inexpensively, completed, and then resent to New York in record time. Even the recent nation-changing political protests, such as the Arab Spring, have been coordinated through the Internet and smart phones to varying extents.

The modern Internet can trace its beginning back to the reaction to the Soviet Union's launch of *Sputnik* in 1957. As one of many responses, the United States created the Advanced Research Projects Agency (ARPA), later called the Defense Advanced Research Projects Agency (DARPA). One goal of the agency was to better facilitate military communication capabilities. ARPA and the Massachusetts Institute of Technology (MIT) created the Advanced Research Projects Agency Network (ARPANET) in 1969, which connected the University of California, Los Angeles, with the Stanford Research Institute (SRI) in Menlo Park, New Jersey. This was subsequently expanded to include more universities. The University of Wisconsin, Madison, created the Computer Science Network (CSNET) in 1981 for universities not on ARPANET. In 1986, the National Science Foundation (NSF), founded by President Truman in the United States in 1950 to promote research and education in the sciences, facilitated a communication network for universities that was more comprehensive, called the National Science Foundation Network (NSFNET). The European Organization for Nuclear Research, also known as CERN, located in Switzerland, opened the World Wide Web (www) network to the public in 1989, which rapidly expanded throughout the 1990s. It is likely that we are witnessing only the beginning of the Internet Revolution that will continue to profoundly influence our lives.

If we think back to the 15th century at the advent of the printing press, people could not have imagined the impact the printing press would have on the world and on the dissemination of knowledge. Furthermore, people could not have fathomed life in the early 21st century. Mathematicians throughout the ages could not have imagined the ease of computations available through even the most basic handheld calculator, let alone CAS and statistical software applications. We are left to speculate what the distant future in technology could bring. It is difficult to say, but perhaps access to a future version of quick information like the Internet, and the ability to compute quickly, could be entirely in our minds. A student of the author of this book believes that people will soon have a computer chip in their brains that gives them instant access to all known information and the ability to compute instantly. In a way, at least from an information perspective, we would all be much smarter than we are now. Perhaps we should take this one step further to say it will not be a computer chip, but rather something else entirely that we have not conceived of yet and perhaps something less invasive than a computer chip. It is also possible that biotechnology might take us to a place where we are able to increase our own intelligence to the point where there will not be a visible line between computer technology and the human mind.

In the next chapter, we focus on the results of one very significant event in the 20th century that led to a major shift in mathematics and science from Europe to North America. In the years prior to World War II, many mathematicians and scientists immigrated to North America, first for better opportunities and second to escape the brutality of Nazi Germany. While the mathematics programs were relatively new compared with the traditional learning centers of Europe, the United States quickly became a center for mathematics and science. The University of Chicago, for example, opened its mathematics department in 1892 and soon started awarding many of the early US doctorates in mathematics. Until the 20th century, most mathematicians in the United States had to conduct their graduate work in Europe, mostly in Germany, but at the turn of the 20th century and onward, American mathematicians could study in the United States and receive a world-class education.

# CHAPTER 16

# EUROPEAN, AMERICAN, AND GLOBAL MATHEMATICS: A SHIFT IN THE 20TH CENTURY

World War II was the largest and deadliest conflict in human history, with approximately 60 million people killed during the war. It proved to have a profound impact on the political and economic outcomes of the remainder of the 20th century and into the 21st century. World War II resulted in the dismantling of the financially and militarily exhausted British Empire. Germany, formerly the leading power in continental Europe, was defeated and occupied by the Western Allies and the Soviet Union. The Soviet Union, despite devastating losses of life with estimates above 25 million people and severe destruction of Soviet cities during the war, emerged as one of two world superpowers, alongside the United States, whose mainland was untouched by war. The United States, the world's leading industrial giant, elevated its position from being a great power to a superpower and was the only major combatant that did not experience the war directly on the homeland.

At the end of World War II, Europe was decisively split into two spheres of influence, the Western American and the Eastern Soviet

*The Development of Mathematics throughout the Centuries: A Brief History in a Cultural Context*, First Edition. Brian R. Evans.
© 2014 John Wiley & Sons, Inc. Published 2014 by John Wiley & Sons, Inc.

divisions. Western powers such as the United Kingdom and France became founding members with the United States of the North Atlantic Treaty Organization (NATO). The Soviet bloc, along with nations such as Poland, Czechoslovakia, and Hungary, made up the Warsaw Pact. The rivalry between the two superpowers was due to differences in ideology, politics, and economic systems. While the United States and its allies were capitalistic democracies, the Soviet Union and its allies were communist totalitarian states. Despite the ideological differences, the rivalry also resulted from global resource competition and a desire for global influence.

World War II began for similar reasons with Adolf Hitler's desire for *Lebensraum*, or living space, which originated from a need for more resources, including farmland and oil. After World War II, more resource competition resulted between the superpowers, as can be found in resource proxy wars as the Vietnam War, which was fought over access to rich resources in Southeast Asia, such as rubber and tin, in addition to ideological influence.

The Cold War between the United States and Soviet Union had its beginnings during and immediately after World War II. Before the end of the war in 1945, there were suspicions between the Western Allies and the Soviet Union. The Soviets thought, perhaps justifiably so, that the Americans and British were delaying the opening of the Western front with an invasion of Nazi-occupied Western Europe. Soviet leader Joseph Stalin thought Soviet forces were being forced to take the full onslaught of the German army, the *Wehrmacht*, which could be alleviated through the Western invasion that did not happen until D-Day on June 6, 1944, about a year before the end of the war in Europe. Rightfully so, the Western Allies did not trust the Soviet Union's intentions for postwar influence in Europe.

Consensus between the Soviets and Western Allies was not reached in negotiations for the political and economic landscape of postwar Europe. The Soviet Union refused to ever allow a nation the opportunity to invade, thus creating a "buffer zone" around its western border by controlling Eastern European puppet regimes on its periphery. Contrary to modern popular belief in the West, the liberation of Western Europe from Nazi control was only a partial victory of World War II. Although Eastern Europe was liberated from the Nazis as well, it was effectively turned over to Soviet control, and one totalitarian regime was traded for another. The United Kingdom and France were interested in maintaining their balance of power and feared a rising Germany at the start of World War II. However, another rising power, the Soviet Union, threw off that balance for them. British Prime Minister Winston Churchill was quick to recognize the scope of what was lost at the end of World War II, but, unfortunately, American Presidents Franklin Roosevelt and Harry Truman did not take action to prevent the Soviet influence in Eastern Europe.

Both the United States and Soviet Union profited greatly from Germany's defeat in gaining access to German technology and scientists. In the 1930s, many scientists and mathematicians immigrated to the United States from Europe for two reasons: the prosperity and opportunity found in the United States through more available academic positions and to escape Nazi Europe. This "brain drain" continued before, during, and after the war. After the war ended, Operation Paperclip was America's effort to bring former Nazi scientists to the United States to apply German technology to American military projects. The Soviets embarked on a similar endeavor called Operation Osoaviakhim. Despite the near certainty of defeat for the Germans in the final years of the war, German science was quite innovative. The world saw the first rocket missiles and jet engines come out of a desperate Germany trying to use modern technology in a last effort to win the war. Both the Americans and Soviets were able to quickly apply advanced German technologies to their own militaries.

The surprise atomic bombing of Japan further led to fear and distrust of the United States by the Soviet Union. A claim could be made that the United States dropped atomic bombs on Hiroshima and Nagasaki as a show of force to the Soviet Union. Another more prominent viewpoint was that the United States used atomic bombs to end the war quickly since Japan would not likely surrender otherwise, thus saving more lives.

By 1945, the US Navy replaced Britain's Royal Navy as the world's most powerful navy and the United States became the world's only nuclear power. At this time, the Soviet Union had the world's greatest land army with more soldiers and tanks than any other nation. On August 29, 1949, the Soviet Union tested its first atomic weapon called *First Lightning*. The American public was shocked because America no longer had the nuclear monopoly. Indeed, the American public knew the Soviet Union was a great threat capable of challenging the United States.

On October 4, 1957, the Soviet Union launched *Sputnik 1*, the world's first artificial satellite sent into orbit. *Sputnik 1* was the first in the *Sputnik* program and signaled the beginning of the Space Race between the Soviet Union and the United States, a logical corollary to the Cold War. To the shocked American public, this signaled that the Soviet Union was at least symbolically ahead in technological advancement, if not realistically ahead. In the minds of many Americans, this proved that the Soviet Union was indeed a more capable foe. Many chose to ignore the fact that the United States maintained nuclear superiority since the advent of nuclear weapons a decade earlier. However, unlike the start of the Atomic Age, in which the United States was in the lead, the Space Age was dominated by the Soviet Union.

The true American concern pertaining to the *Sputnik* launch was regarding military applications. Although *Sputnik* may have appeared to be harmless, a small ball-like sphere orbiting the Earth making an incessant beeping sound, the real problem was that it was providing the Soviets with valuable information about the atmospheric density that would enable them to develop better trajectories for their missile program.

The United States responded by quickly launching its own satellite into orbit, *Explorer 1*, which was the first American artificial satellite, sent into orbit on January 31, 1958. However, other Soviet accomplishments such as sending the first human, Yuri Gagarin, into orbit on April 12, 1961, indicated that the Soviets may have been pulling ahead of the Americans yet again. The United States reacted by sending Alan Shepard, the second human, into orbit on May 5, 1961. It was not until July 20, 1969, when American astronaut Neil Armstrong became the first human to walk on the Moon that the United States was able to claim a substantial lead in the Space Race.

The early Soviet lead in the Space Race prompted US President Dwight Eisenhower to create the National Aeronautics and Space Administration (NASA), as well as to improve American education in mathematics and science. This led to the National Defense Education Act (NDEA), passed by Congress in 1958. The NDEA provided support in critical school subjects such as mathematics, science, and foreign languages. Currently, groups such as the Association of American Universities call for Congress to enact a new NDEA for the 21st century. In 1958, the American Mathematical Society (AMS), a mathematical association founded in New York in 1888, formed the School Mathematics Study Group (SMSG) to develop a new high school curriculum. In his January 25, 2011, State of the Union address, President Barack Obama claimed that the United States is at another "*Sputnik* moment" in which the United States is falling behind other countries, especially the countries whose students achieve highly on international mathematics and science testing.

A major reform that occurred in mathematics education was the advent of "New Math." New Math, which was influenced by the work of the Bourbaki mathematicians, emphasized the study of axiomatic systems, set theory, and base systems outside base 10. This new approach was confusing for students, parents, and teachers and was highly criticized for its lack of traditional instruction on basics such as computation and memorization of the multiplication tables. New Math was considered to be highly ineffective by almost all educators and was often the object of jokes in popular culture.

Despite the efforts of New Math, the nation was in an educational, and perceived technological and military, crisis. American children were not learning enough mathematics or science to compete with America's Cold War foe. The perception of crisis in the United States subsided with the success of the American space program over the next decade. Even though the Space Race was a race in technological superiority, the foundation of the Space Race is a superior mathematics and science education program starting in the early years of children's educations. If the Cold War technology competition between the world's two superpowers can be labeled the Space Race, the current crisis we face is the Globalization Race. Both the Space and Globalization races are subcategories of the Mathematics and Technology Race since mathematics is the foundation of scientific and technological innovation.

Today, the United States is still the world's leading economic and military power, especially after the dissolution of the Soviet Union in 1991. Following the aftermath of World War II in 1945, many of the world's nations have long been revitalizing their economies, as seen in Germany and Japan in the late 20th century. The 20th century has been referred to as the American Century, just as the 19th century was called the British Century due to the dominance of the British Empire. During the 20th century, the United States accumulated enormous wealth and had many poor citizens shift into the middle class. The rate of industrialization was remarkable. The US victory in the Spanish-American War in 1898 marked the beginning of things to come for the young and increasingly prosperous nation. Not only did the United States acquire the Spanish colonies of Cuba, Guam, the Philippines, and Puerto Rico but it also entered the global stage as a world power on par with other great European powers.

If it was the Spanish-American War that placed the United States on the world stage, it was the conclusion of World War II that solidified that. The traditional great powers of the modern world, the United Kingdom, France, Germany, Italy, and Japan, were exhausted and devastated from war. The only great power that did not have its infrastructure badly damaged was the United States. The United States and the Soviet Union rose as the world's two great superpowers at the end of World War II and remained so in varying degrees during the Cold War and throughout most of the 20th century. The once mighty British Empire barely avoided bankruptcy in 1946 and watched the empire break apart over the next several decades.

It is common in the news headlines today to read about the Chinese Century or the Asian Century. Two hundred years ago, Napoleon said that we should let China sleep, for when China wakes, she will shake the world. It is quite clear that China has awakened and is asserting herself on the world's economic stage. Today, the United States is being challenged by rising economic powers such as China and India. Russia, the former Cold War arch foe, is among the list of rising economic superpowers with its vast energy supplies of oil and natural gas. Russia is increasingly becoming a global, economic, and political force in the world. Goldman Sachs, one of the world's largest investment banks, used the term BRIC to encompass the four nations of Brazil, Russia, India, and China as emerging economic superpowers heading toward the mid-21st century. China and India are regarded as manufacturing and service economies, respectively, while Russia and Brazil are regarding as raw material economies.

Thomas Friedman, in his book *The World Is Flat: A Brief History of the 21st Century*, contended that there are many factors that are leveling the playing field in economic international competition. It is in our best interest to focus on two "flatteners." First is the collapse of communism. Two events that best symbolize this collapse were the fall of the Berlin Wall and the demise of the Soviet Union. This opened new markets to global capitalism, such as India, a former Soviet ally, and Eastern Europe. Second, around the same time, another major shift occurred with the beginning of the Information Age. It is difficult to date the birth of the Information Age, but it might make sense to use 1989, the year the World Wide Web was introduced. Even though it was not until the later 1990s that the Internet became ubiquitous in the United States, it is important to note that 1989 was the same year the Berlin Wall collapsed. Remarkably, these two factors seem to have worked simultaneously. By the early 21st century, the Internet became widely accessible to a large percentage of the population in the Western world, and increasingly so in other parts of the world.

In a postindustrialized society such as the United States, scientific and technological advancement is a major contributor to economic success, which supports national security. In a postindustrialized world, the basis for an economically and military advanced nation is a populace that is competitively advanced in mathematics and science. There is no other investment that has the potential for higher returns than an investment in mathematics and science education.

In the next chapter, we examine the history of mathematics education in the United States. We shall explore the evolution of mathematics education from early education in the United States to the current debates about the best route to take and methods to use in educating children in mathematics.

# CHAPTER 17

# AMERICAN MATHEMATICS EDUCATION: THE 20TH AND 21ST CENTURIES

The history of education in the United States begins in the 17th century with the establishment of Boston Latin School in 1635 as the first school in the United States. The following year, the New College, which became Harvard University, was founded in Massachusetts, and shortly after, there was the establishment of grammar schools that were supported by tuition in Massachusetts, Connecticut, New York, and Pennsylvania. In 1693, the College of William and Mary was established in Virginia. In 1821, the first public high school was established in Boston. Thomas Jefferson, the third president of the United States, believed that an educated populace was needed to secure a functioning democracy. Jefferson promoted the idea of a free education for all, and was instrumental in the establishment of the first state university in the United States, the University of Virginia in 1825.

Public education spread rapidly in the early 19th century with the work of Horace Mann, who was an education proponent born in 1796 in Massachusetts and is referred to as the "Father of Public Education." In 1837, Mann was appointed as Secretary of the Massachusetts State Board of Education, and this began the period known as the Common School Movement. Common schools were established to

*The Development of Mathematics throughout the Centuries: A Brief History in a Cultural Context*, First Edition. Brian R. Evans.
© 2014 John Wiley & Sons, Inc. Published 2014 by John Wiley & Sons, Inc.

educate children from all social and economic classes. In 1839, Mann established a normal school to train teachers in Massachusetts. Henry Barnard, the Secretary of the Connecticut State Board of Education, was also an early proponent of the Common School Movement and continued Mann's work in Connecticut. Compulsory education spread all over the United States throughout the century. Mann died in 1859.

John Dewey, an education reformer born in Vermont in 1859, the same year Mann died, was at the head of the Progressive Movement in education that spanned the early part of the 20th century. He is called the "Father of Modern Education." Dewey received his doctorate from Johns Hopkins University. He held positions at the University of Chicago from 1894 to 1904 and Columbia University from 1904 to 1930 when he retired. The Progressive Movement emphasized student learning through active student engagement and through social interaction. Students approached their learning through the scientific method, and the ultimate goal was student understanding of the content. Dewey died in 1952.

Around the same time John Dewey was active in education, Edward Thorndike's work was also notable, which is the reason he is often referred to as the "Father of Educational Psychology." Thorndike was born in Massachusetts in 1874. He received his doctorate from Columbia University in 1898 and spent most of his career there. Thorndike was an early pioneer in measuring student achievement and was influential in the current interest in standardized testing. He created intelligence tests for the US Army in World War I in order to better sort soldiers for duties that varied based on intelligence. Thorndike died in 1949.

In 1923, a report was published by the Mathematical Association of America (MAA), a mathematical association founded in 1915 to promote mathematics and learning. This report was *The Reorganization of Mathematics in Secondary Education* and it was concerned with the quality of mathematics education in the United States based on the reorganization of mathematics courses and focused on improving teaching. It was a response to William Heard Kilpatrick's *The Problem of Mathematics in Secondary Education* report in 1920. Kilpatrick was part of the Progressive Movement and a successor to Dewey. He proposed less emphasis on mathematics for certain students. We can consider this an early version of the Math Wars that ensued toward the end of the century.

In 1925, a trial was undertaken to test the teaching of Darwinian evolution in the classroom. This trial was the *State of Tennessee v. Scopes*, but was later known as the Scopes Monkey Trial. John Scopes, a high school biology teacher, was accused of violating the Butler Act, which forbade the teaching of evolution. Former presidential candidate and politician William Jennings Bryan represented the prosecution, and agnostic lawyer and member of the American Civil Liberties Union, Clarence Darrow, represented Scopes' defense. Scopes lost the trial and was fined US$100, but the trial brought to the public's attention the debate between evolution and creationism and was considered a victory for the proponents of evolution. Unfortunately, even though the theory of evolution is accepted in the scientific community, this debate surfaces all too often today.

At the outbreak of World War II, a major focus was placed on preparing citizens who would be capable of performing the basic mathematical tasks needed in war. Soldiers needed to be mathematically literate in order to do their jobs. As discussed in previous chapters, World War II made it clear that mathematics and science were important factors in winning wars. By the 1950s, around the time of Dewey's death, the Progressive Movement was criticized and began to wane in influence. In 1955, Rudolf Flesch's famous book, *Why Johnny Can't Read*, was published. This book criticized the approaches taken by the Progressive Movement.

George Pólya, a Hungarian mathematician born in 1887 in Budapest, became a significant figure in 20th-century mathematics education. Pólya's family was Jewish, but had converted to Catholicism before he was born. Although a bright student, Pólya did not do very well in mathematics class nor did he particularly enjoy mathematics during his childhood. Pólya later described the quality of his mathematics education as poor. At the university, Pólya became interested in philosophy, and it was recommended that he study mathematics to better understand philosophy. When asked why so many great mathematicians originated in Hungary, a relatively poor country, Pólya famously said that mathematics is cheaper than the other sciences since the only equipment needed is a pencil and paper. In 1913, while Pólya was at the University of Göttingen, he was taking a train and became involved in an altercation with another man, who happened to be a student at the University of Göttingen. Because of this, Pólya had to leave and took a position in Switzerland. At the outbreak of World War II, Pólya left Europe for the United States and eventually took a position at Stanford University until his retirement in 1953. Pólya lived a long life and died at 1985 at 97 years old.

While Pólya did very significant work in many areas of mathematics including analysis, probability, and geometry, among others, it is his work in mathematics education that may be most remembered. Póyla's interest was to make mathematical problem solving more systematic and understandable for students. When many people think of problem solving, the first thing that may come to mind is the typical word problem. While a word problem can involve a problem-solving process, it is important to realize that not all word problems use true problem solving. If a teacher gives a student a word problem, but has already modeled how to find the solution to a similar word problem, then this is not problem solving. Problem solving involves a student approaching an unfamiliar mathematics problem situation in which the solution is not obvious and the student is required to rely on prior knowledge. The student must approach the problem using critical thinking and reasoning. Pólya said that if a person cannot find a solution to a problem, it might be easier to first find the solution to a simpler problem. In 1945, Pólya published a book called *How to Solve It*, which outlined the steps necessary for systematic mathematical problem solving analogous to the scientific method. Pólya presented a four-step approach.

1. Understand the problem.
2. Devise a plan.

3. Carry out the plan.
4. Look back.

Pólya also presented the reader with multiple strategies for problem solving. Some examples follow.

1. Look for a pattern.
2. Work backward.
3. Guess and check.
4. Draw a picture.
5. Orderly listing.
6. Use a model.
7. Solve a simpler problem.

Today, there is great emphasis on having students learn mathematics through problem solving. Educational research supports that students learn mathematics best through problem solving, and many mathematics education organizations promote learning mathematics through problem solving. Moreover, there is great effort in teacher preparation programs to have teachers learn how to teach mathematics from a problem-solving approach. Not only will students benefit in their learning of mathematics by engaging in problem solving but they will also increase their capacity for critical thought as well.

In the previous chapter, the consequences for the United States from the Soviet Union's launch of *Sputnik* were addressed. One consequence was reform in mathematics education with the implementation of "New Math" into American schools throughout the 1960s. As said in the previous chapter, this new curriculum was considered by many to have been a failure. In an attempt to treat students as mathematicians, much of the work that many considered basic to mathematics was not addressed. In the 1960s, Tom Lehrer, a mathematician born in New York in 1928, wrote a popular song called "New Math" in which Lehrer lampooned the New Math curriculum through criticism of computations outside base 10 and the struggle many adults and parents had with the content learned by children.

In 1973, Morris Kline, a mathematician born in New York in 1908 who was a professor at New York University, published a book called *Why Johnny Can't Add: The Failure of the New Math*. In his book, Kline criticized the educational reforms of New Math, and this ushered in the "Back to Basics" movement in the 1970s with an emphasis on proficiency with computation, number operations, and memorization of important facts, such as the multiplication tables. Klein died in 1992.

The National Council of Teachers of Mathematics (NCTM), an organization for mathematics teachers founded in 1920, has become an influential voice in the direction of mathematics education. In 1980, NCTM released the document *An Agenda for Action*. This document emphasized problem solving in the mathematics curriculum and advocated for technology to be used in instruction, such as student use of the now widely available and inexpensive calculators.

Shortly after *An Agenda for Action* was released, President Ronald Reagan's National Commission on Excellence in Education released a report in 1983 called *A Nation at Risk: The Imperative for Educational Reform*. The report focused on education in general, not only mathematics. For mathematics, it recommended that students take more courses with more rigorous content, and it addressed the issue of teacher shortages in mathematics and science by proposing the recruitment of highly qualified people to teach, and to pay them well to do so. *A Nation at Risk* was an early report to emphasize standardized testing, an issue that has been greatly debated since.

In 1989, NCTM released its *Curriculum and Evaluation Standards for School Mathematics*, which propelled mathematics education into the standards movement and reform-based instruction of the 1990s and onward. The document emphasized many of the issues in *An Agenda for Action*, such as an emphasis on problem solving. NCTM proposed that instruction should be student-centered, just as Dewey and the Progressive Movement had recommended. NCTM focused on conceptual understanding instead of the traditional focus on procedural knowledge. A constructivist approach was taken, which is a school of educational thought promoted by thinkers such as Jean Piaget and Jerome Bruner, which advocated having students construct or develop their own knowledge with the teacher acting as a guide or coach instead of giving direct instruction. This would allow for a guided inquiry approach to education, which would give students the opportunity to explore the mathematics on their own.

The 1989 release of *Curriculum and Evaluation Standards for School Mathematics* could be considered the beginning of the "Math Wars" in the late 20th and early 21st centuries. On one side, there is the NCTM with its reformed-based approach, and on the other side are those who advocate a traditional approach to learning including direct instruction, memorization, and the use of algorithms, led by groups such as Honest and Open Logical Debate (HOLD) and Mathematically Correct, among others. The traditionalists believe that students must master the basics before moving on to conceptual understanding, while the reformists believe that students will learn the basics when they explore deep mathematical concepts. Traditionalists want students to learn the traditional methods of computation, such as the traditional method of multiplication, while reformists want multiple approaches, such as exploring the Egyptian method and the lattice methods of multiplication, as detailed earlier in the book. The use of technology, such as calculators, is fiercely debated between both sides. The debate can be narrowly defined as a debate on teaching method since both sides have the goal of mathematical competency for students.

In 2000, NCTM updated its standards document with the release of the *Principles and Standards for School Mathematics*, also known as the *Standards*, which emphasized mathematics education from preschool to grade 12. The document was written with input from educational researchers, mathematicians, and teachers, and it came to influence individual state standards throughout the United States. It was seen by many as taking both sides of the debate into consideration, although still remaining on the reformist side. Hence, the Math Wars continued. This document focused on six principles in mathematics education and ten standards, which could be divided into five content and five process standards. The six principles are equity, curriculum, teaching, learning, assessment, and technology. NCTM promoted treating students with equity, which means giving accommodations to students who need it. The curriculum needs to be coherent and connected, and to emphasize important mathematics. Teaching and learning should focus on supporting student learning and understanding mathematics while building on previously acquired knowledge. Assessment should not only produce a final grade for students but also help teachers guide instruction and give vital information to students and parents in areas in need of more development. Finally, technology influences the mathematics that is taught and how it is taught, and supports instruction.

The five content standards are what students need to know to include number and operations, algebra, geometry, measurement, and data analysis and probability. The five process standards are how students learn the content and include problem solving, reasoning and proof, communication, connections, and representations. Students must be able to communicate their mathematical knowledge and must have multiple ways of representing it. Connections mean that students see mathematics connected to other subjects and the real world, as well as the interconnectedness of different areas of mathematics, such as algebra and geometry.

The Trends in International Mathematics and Science Study (TIMSS) is an international assessment of fourth and eighth grade students' achievement in mathematics and science content knowledge that has been conducted every 4 years beginning in 1995. While the United States has been improving its rankings over the subsequent studies, it continues to rank lower among participating nations than many in the United States would prefer. The highest scoring nations mostly come from the countries and administrative regions of East Asia such as Singapore, Hong Kong, Japan, South Korea, and Taiwan. In 2001, the US Commission on National Security in the Twenty-First Century reported that inadequacies of US education pose one of the biggest threats to the national security of the United

States. The dismal scores of US students on the 1995 TIMSS report could be considered another *Sputnik*-like moment in US history. Liping Ma's 1999 influential work, *Knowing and Teaching Elementary Mathematics*, exposed the differences in conceptual understanding of elementary school teachers in the United States and China. In many cases, the Chinese teachers, with less teacher preparation, were able to outperform their US counterparts, who had more preparation. Interestingly, Ma's book was promoted on both sides of the Math Wars.

After the initial TIMSS reports were released, the traditionalist side in the Math Wars claimed that this was evidence that the reformist approach did not work. The reformist response was that the recommendations they made were not being fully carried out. For example, many teachers claim to favor reformed-based methods of instruction, but on analysis, it is clear that they practice traditional methods. The traditionalists pointed out that Singapore, the highest achieving nation in most cycles of the TIMSS study, used methods that appeared traditional. At the turn of the 21st century, a publishing company, Singapore Math Inc., published mathematics books based on the Singapore method of mathematics instruction in the United States. These books were based on the mathematics textbooks used in Singapore, but adapted for a US audience. These books, along with books by publishers such as Saxon, have been embraced by the traditionalists. The reformists promote books such as *Connected Mathematics* and *Everyday Mathematics*, referenced in Chapters 2 and 8. Interestingly, the Singapore mathematics books appear to have features of reform-based mathematics, such as emphasis on conceptual understanding and problem solving, and they are much briefer than US books and cover fewer topics more deeply, an idea promoted by the reformists. A common criticism of reform-based mathematics books and curricula is that many teachers and parents do not know or understand the methods being used, which means that it makes teaching the material more difficult for the teachers and prohibits many parents from helping their children with homework. It is imperative that if reformed-based materials are being used in the classrooms, that both teachers and parents receive the support they need to understand the material and methods in order to best serve the children.

In 2001, President George W. Bush signed the No Child Left Behind Act (NCLB) into legislation, which placed emphasis on standards-based education in all subject areas and standardized testing to assess student proficiency in meeting those standards. This required teachers to be highly qualified and ushered in an age of accountability. There has been deep criticism of NCLB because of its emphasis on standardized testing. Critics believe that it forces teachers to "teach to the test" due to the consequences for schools, such as reduced federal funding, that do not have well-performing students. This comes at the expense of real learning for understanding. Another criticism is the unrealistic goals NCLB places on the students and schools with little support from the government.

In 2006, NCTM released its *Curriculum Focal Points for Prekindergarten through Grade 8 Mathematics: A Quest for Coherence*, which appeared to further alleviate the Math Wars by placing traditional algorithms into the curriculum. NCTM outlined the most important mathematics to be learned from preschool to grade 8, and the *Curriculum Focal Points* supported the NCTM *Standards*. A common criticism of the curricula in the United States is the large number of topics covered. The *Curriculum Focal Points* emphasized the most important mathematics to be learned.

Another development in 2006 was the creation of the National Mathematics Advisory Panel by President Bush with the intention of examining the research in mathematics education teaching and learning. The conclusion, published in 2008, was that having a completely student-centered or teacher-centered approach to mathematics education was not optimal. In other words, having students completely discover mathematics on their own or having teachers always provide instruction through direct lecture was not the best approach to take. The panel recommended the middle path to teaching and learning, based on the current research. It called for an end to the Math Wars.

In 2009, President Obama announced the US Department of Education Race to the Top program, in which states that implement reforms into their state education systems could receive a part of over US$4 billion. Criteria included improving teacher effectiveness, improving low-achieving student test scores, and implementing the new Common Core State Standards, which were developed by National Governors Association Center for Best Practices and the Council of Chief State Schools Officers in 2009. In the United States, each state is responsible for its own standards and board of education. However, this initiative seeks to standardize education throughout the United States. Considering the deep economic crisis at the end of the first decade in the 21st century, granting federal funding to financially struggling states to make these reforms has been a strong incentive that has been embraced by many states. Each state is permitted to include up to 15% of their own standards in the Common Core, but there is 85% consistency between the states adopting the new standards. The mathematics standards include mathematical practice and content, which are similar to the NCTM process and content standards. Most of the states have adopted the Common Core Standards.

There have been many efforts to direct the public's attention toward mathematics. John Paulos, a mathematician born in 1945 in Chicago who writes popular mathematics books, said in his 1988 book, *Innumeracy: Mathematical Illiteracy and Its Consequences*, that the general public has a negative attitude toward mathematics and a profound misunderstanding of the subject. Other studies have shown that there is a relationship between attitude toward mathematics and achievement in mathematics. Paulos claimed that there is a misdirected pride among the general public in their lack of mathematical understanding. Mathematics teachers need to find innovative ways to present the material to students to gain their interest. One method favored by the author is to incorporate mathematics history into the classes to show the human side of mathematics (which may make a book like this one very useful for teachers). One such effort to increase interest in mathematics is the celebration in schools of Pi Day, founded by Larry Shaw in 1989. Pi Day takes place on March 14, the day before the Ides of March, which was the death of Julius Caesar in 44 BCE.

However, Pi Day has nothing to do with Caesar. Instead, March 14 is important because it can be written as 3/14, which indicates the approximation of $\pi$ as 3.14. Pi minute is 1:59 am or pm, with pm preferred for the school day, since a better approximation of $\pi$ is 3.14159. We can even go as far as pi seconds as 1:59 and 26 seconds could represent $\pi$'s approximation of 3.1415926. Pi Approximation Day is July 22, a day not included in the standard school year, because 22/7 is a popular approximation of $\pi$. In 2009, the US Congress official recognized Pi Day as National Pi Day. Pi Day is celebrated in some schools with eating pie and discussing mathematics.

Mathematics education will continue to evolve over the course of the 21st century. Technology will inevitably affect the way we live and learn. Just as the Internet allows a student instant access to historical facts so that social studies education can focus more on critical issues and historical analysis, mathematics technology such as CAS, statistical software, and graphing calculators will continue to make problem solving and critical thinking important matters for the mathematics classroom. While we should be certain that students have an understanding of the basic facts of mathematics, we have to balance that need with the changing needs of society and the tools students will have available to them throughout the 21st century.

# RESOURCES AND RECOMMENDED READINGS

The following resources were consulted in preparation of this book. These resources are recommended for the reader interested in more detail on the history of mathematics. The MacTutor History of Mathematics is a particularly valuable resource for a comprehensive overview on many people and topics in mathematics history. Klein (2003) and Schoenfeld (2004) provide insight on history of mathematics education in the United States, and Tehie (2007) provides a history of education.

Baumgart, J. K. (1989). *Historical Topics for the Mathematics Classroom*. Reston, VA: National Council of Teachers of Mathematics.

Berlinghoff, W. P., and Gouvea, F. Q. (2003). *Math through the Ages: A Gentle History for Teachers and Others*. Farmington, ME: Oxton House Publishers.

Boyer, C. B., and Merzbach, U. C. (2011). *A History of Mathematics* (3rd ed.). Hoboken, NJ: Wiley.

Calinger, R. (1999). *A Contextual History of Mathematics*. Upper Saddle River, NJ: Prentice Hall.

Cooke, R. L. (2012). *The History of Mathematics: A Brief Course* (3rd ed.). Hoboken, NJ: Wiley.

*The Development of Mathematics throughout the Centuries: A Brief History in a Cultural Context*, First Edition. Brian R. Evans.
© 2014 John Wiley & Sons, Inc. Published 2014 by John Wiley & Sons, Inc.

Du Sautoy, M. (2006). *The Story of Maths* (Television Series). British Broadcasting Corporation.

Gullberg, J., and Hilton, P. (1997). *Mathematics: From the Birth of Numbers*. New York: W. W. Norton & Company.

Katz, V. J. (2004). *A History of Mathematics: Brief Edition*. Boston, MA: Addison Wesley.

Katz, V. J. (2008). *History of Mathematics* (3rd ed.). Boston, MA: Addison Wesley.

Klein, D. (2003). A brief history of American K-12 mathematics education in the 20th century. Retrieved from http://www.csun.edu/~vcmth00m/AHistory.html.

Lewinter, M., and Widulski, W. (2002). *The Saga of Mathematics: A Brief History*. Upper Saddle River, NJ: Prentice Hall.

O'Connor, J. J., and Robertson, E. F. (2013). *MacTutor History of Mathematics Archive*. Retrieved from http://www-history.mcs.st-and.ac.uk/history/.

Pickover, C. A. (2012). *The Math Book: From Pythagoras to the 57th Dimension, 250 Milestones in the History of Mathematics*. New York: Sterling.

Schoenfeld, A. H. (2004). The math wars. *Educational Policy, 18*(1), 253–286.

Smith, S. M. (1996). *Agnesi to Zeno: Over 100 Vignettes from the History of Math*. Emeryville, CA: Key Curriculum Press.

Stillwell, J. (2010). *Mathematics and Its History*. New York: Springer.

Struik, D. J. (1987). *A Concise History of Mathematics* (4th ed.). Mineola, NY: Dover Publications.

Tehie, J. B. (2007). *Historical Foundations of Education: Bridges from the Ancient World to the Present*. Upper Saddle River, NJ: Merrill/Prentice Hall.

# INDEX

Printed and bound by CPI Group (UK) Ltd, Croydon, CR0 4YY

27/10/2024

14580265-0002